MACHINE PATTERN MAKING

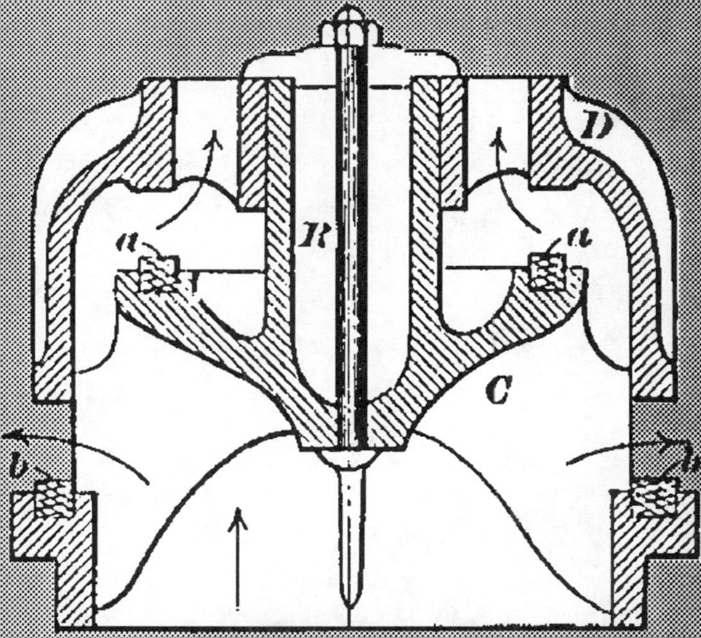

P. S. Dingey

Machinery Pattern Making

CONTAINING

FULL SIZE PROFILES OF GEAR TEETH

AND

Fine Engravings on Full-Page Plates, Illustrating Manner of Constructing Numerous and Important Patterns and Core Boxes

BY

P. S. DINGEY

ASSOCIATE MEMBER OF AMERICAN SOCIETY MECHANICAL ENGINEERS

417 Fine Illustrations

SECOND EDITION. REVISED AND ENLARGED

FIRST THOUSAND

Machinery Pattern Making

by P. S. Dingey

Originally published by
John Wiley & Sons Inc
New York

Reprinted by Lindsay Publications Inc
Bradley IL 60915

ISBN 1-55918-039-0

1 2 3 4 5 6 7 8 9 0

1990

NOTE.

Most of the matter in this book was written expressly for the *American Machinist,* to whose courtesy we are indebted for some of the illustrations. Mr. Dingey has, however, revised many of his drawings and much of the matter, adding some valuable items.

WILEY & SONS.

WARNING

Remember that the materials and methods described here are from another era. Workers were less safety conscious then, and some methods may be downright dangerous. Be careful! Use good solid judgement in your work, and think ahead. Lindsay Publications Inc. has not tested these methods and materials and does not endorse them. Our job is merely to pass along to you information from another era. Safety is your responsibility.

Write for a complete catalog of unusual books available from:

Lindsay Publications Inc
PO Box 12
Bradley IL 60915-0012

PREFACE TO THE SECOND EDITION.

In issuing this edition the author wishes to thank the pattern-makers, molders, and those in the machinery business and colleges for the appreciation given to the first edition, and trusts that the second will receive a similar patronage and consideration. It is hoped that among the pages added, those on double-beat valves, the propeller, governor valves, plug valves, and items for pattern-makers will increase the interest already shown.

The object of this book is not so much to teach pattern-making—for that cannot be done alone through a book—but merely to encourage a study of the practical.

No cast-iron rules have been laid down, and it is desired that the contents be accepted as suggestions; at the same time that which is given is practicable, being the result of practice and of over twenty years' experience in the business.

In the present volume everything of a visionary kind has been avoided, and the author has presented such subjects as he believes will be interesting to pattern-makers and to the machinery business generally.

P. S. Dingey.

Chicago, Ill., October 30, 1894.

CONTENTS.

CONTENTS.

THE PATTERN MAKER AND HIS TRADE.

For one to be a good pattern maker it is indispensable that he should have a theoretical knowledge of moulding, and be able to read drawings readily—for he MUST know how a pattern is to be moulded, and to mentally see the machine, or parts of it, just as the draftsman does before he can do much toward making the pattern.

To become an expert in handling tools is the most necessary requirement in many trades, but this is not the case in pattern making—there being something far more important than cutting wood.

In very many patterns it is not so much a question of workmanship as of knowledge, for there are certain patterns which, after they are made, may be duplicated by any good wood worker, workmanship not having been the all-important; in fact, some very important jobs require but very little skill to construct the necessary core-boxes, etc., that may be required to produce certain castings. On the other hand there is much that calls for fine workmanship and less scheming. This is no doubt true, more or less, in all trades, but it is especially so in pattern making, and this is why I say that pattern making is not merely cutting wood. From the very nature of the trade, a pattern maker is a good worker in wood, because he is accustomed to work to finer measurements than the ordinary wood worker.

The pattern maker's position in some establishments is an important one, and the responsibility resting upon the leading man or foreman of that department, as to whether work turns out right, is equal to that of the drawing room; for while the draftsman is responsible for the design, upon the pattern maker rests a large proportion of the responsibility of executing correctly that which has been put upon paper.

The liability to mistakes is reduced considerably when the machinist takes hold where the pattern maker has left off; the machinist's part is no doubt the most important as to the workmanship and right working of the machinery; he can make it good, bad, or indifferent; but mistakes in measurements he is not so liable to as the pattern maker, because the machinist has the casting, and is given the drawing of it with instructions to finish to drawing.

When a pattern maker is given a drawing, he has to imagine the casting before him, and build something that will produce it; it may be called a pattern, but often it is really not a pattern of what is wanted, it being all core-boxes and no pattern, and here is where the responsibility comes in, and will, I think, explain why the pattern shop is often the birth-place of mistakes.

Of course, mistakes ought not to occur; but as long as pattern makers are fallible, they will happen sometimes, though the utmost precaution be taken. I am always suspicious of the man that never makes mistakes; he will need watching, for the over-confident man is not to be trusted any more than those careless pattern makers who are constantly making blunders, and who think when their patterns come within an eighth of an inch it is near enough.

From the nature of the trade of machinery pattern

making, there is more danger of errors being made in that branch of machinery building than others, this being so, the careful, industrious workman, who seldom makes an error, is worthy of consideration when he does happen to be caught, for such a man usually feels bad enough over his mistakes, without having the foreman make him feel worse.

Owing to the advance made in mechanical arts, pattern making is becoming one of the most important branches in machinery building. It is often underrated by a class of machinists who think that because a pattern maker is not called upon to work in iron, and to one-hundredth or one-thousandth part of an inch, that there is not much in pattern making; and yet the pattern maker is as much of a machinist, in reality, as those generally known as such.

The onward march of improvements in machinery demands that the pattern maker must keep right up abreast with the times, although he is considered "a necessary evil" among manufacturers.

There is a great deal of machinery now constructed, the coring of which is so complicated that it taxes the ingenuity of both pattern maker and moulder to know how it can be made at all—the winding passages and secret chambers that are wanted in some castings, are worse than those we read about in books. The old fashioned idea of bolting on an arm here, and screwing on a bracket there, are fast disappearing. The modern plan is to make a machine with as few pieces as possible, thus making the pattern more difficult to build.

There are many patterns that require little or no

knowledge of pattern making to make, but I would not advise anyone, because he has made a few such patterns, to pose as a pattern maker; there are those who do. I have had some experience with them, and hope always to be delivered from such. They are a worry to any foreman—he is in constant fear that with all his watching, the would-be pattern maker will make some serious blunder that will cost the firm a considerable sum of money—for a mistake in the pattern means a mistake in the casting, and as an old employer of mine used to say: "Cast iron mistakes are rather serious things."

The fact that there are so many different ways of moulding, gives a great field for study for the pattern maker, as to the best way of making a pattern; but whenever a complicated piece of work is to be done, the moulder should be consulted, and I do not think that the pattern maker will lose any of his ideas by consulting with his brother, the moulder, and while the practical parts of the two trades are as unlike as possible, yet there is a connection between the moulder and the pattern maker that is inseparable. If discussion is necessary, let it be carried on intelligently, each respecting the other's opinions. Wherever this is done, good is sure to result, and the chances are that the best way of doing a job will be arrived at. There are those who are so eager to advance their own ideas and have them carried out, that they are unwilling to consider those of others; such persons are not likely to be very profitable to any concern, for they think more of airing their own genius than of arriving at any results that might be of practical value.

THE PATTERN SHOP.

ITS POSITION, SIZE, AND REQUIREMENTS.

THE question has often occurred to me why pattern shops should be located on the upper floor of a building, as they usually are, where the foundation—if it may be termed such—for fast-running machinery is anything but what it should be. It is sometimes risky business turning a large pattern in a lathe whose only foundation is an upper floor that springs with every motion of the machinery, and the chances that pattern makers take when turning a large pattern under these conditions are great.

The foundation for large face lathes for turning large diameters cannot be too rigid, but I have known the trembling of a large face plate to be caused not so much by the floor as by having an arbor that was too small, or bearings that were set too closely together.

The inconvenience attending the getting of large patterns up and down a stairway, or on an elevator, as the case may be, is another reason why the ground floor is a better location for a pattern shop.

Plenty of room and light are also essentials generally lost sight of. The pattern shop is sometimes called the pattern *room*, and with many firms it has been rightly named, having, as is often the case, such a small space set apart for that purpose that it has scarcely deserved the name of shop.

The nature of the trade that is carried on, determines, in a large measure, the size pattern shop a firm requires.

A firm of large dimensions making specialties does not need such a large pattern shop as a smaller one that builds engines and general machinery. It is more to this latter class of manufactories, employing about twelve to fourteen pattern makers, that reference is made.

Enter a number of manufacturing concerns, and in nine out of ten, it will be found that the pattern makers are working so closely together as to prevent them from getting around their work in a proper manner; and it is surprising how a job may be impeded for lack of room to build it.

When it happens that there is a run of large work, then it is that the oft repeated expression is heard "We ought to have a larger pattern shop." It is true that shops are sometimes situated on such valuable property that a limited space only can be allotted to each department; but this does not do away with the fact mentioned. In these days of sharp competition, the firms that are not cramped for space are the successful competitors in the machinery business.

A pattern shop about 75 ft. x 50 would be a convenient size for working the number of men named.

The machines should not be located all over the shop, but at one end within a reasonable working distance of each other.

Among the requirements of such a shop would be a face lathe for turning large patterns, 30″ lathe with bed about 18 ft. long, 16″ lathe for small work, combination circular saw table, plain saw table, with saw about 12″ diameter, band saw, jig saw, surface planer, Daniel's planer, two or three Fox Trimmers, and about six dozen (rather more than less) of assorted clamps. I mention

Fig. 1.

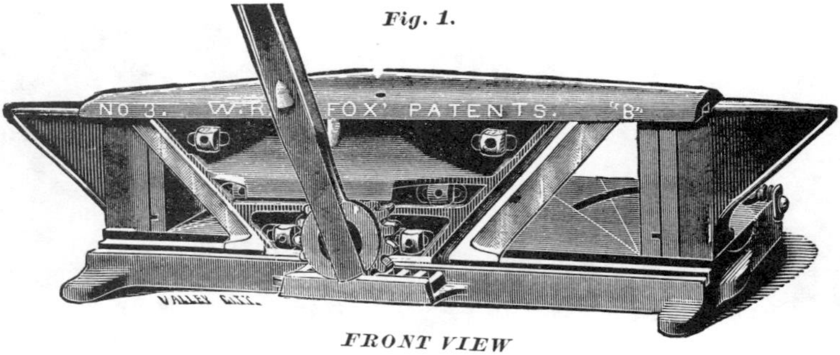

FRONT VIEW

Fig. 2.

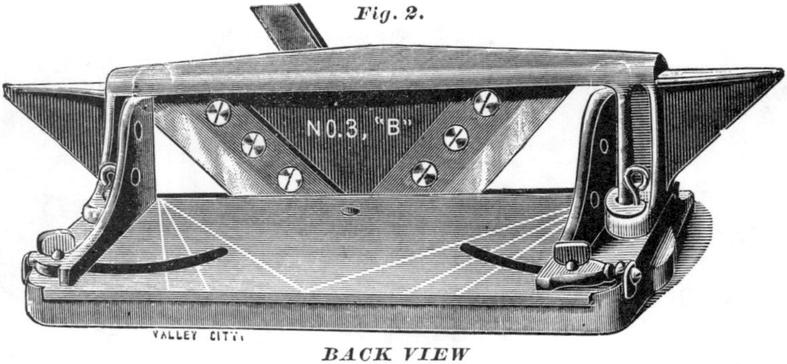

BACK VIEW

6 *in. STROKE:* 3 *in. RISE:* *BED* 6 x 15 *inches:* *Weight* 25 *lbs:*
For SMALL PATTERN WORK.

FOX'S TRIMMER.

this smaller item of clamps in order to insure a plentiful supply. Much time is frequently lost by men waiting on each other for clamps.

The Daniel's Planer is a machine that no pattern shop of any pretentions should be without. For surfacing stuff for pattern makers this machine has no equal, especially when knives are kept sharp, and a good supply should always be on hand.

The Trimmer mentioned is also a very valuable addition to the pattern shop, in fact, it has come to be a standard tool, and the shop that is without one is away behind and had better hurry up and get at least one.

I believe the success of this machine and its being adopted so generally in pattern making, is due to the fact that it was invented by a pattern maker who designed it at the time for pattern making. The Trimmer has certainly done away with a great deal of the paring that used to done with a chisel, and which was exceedingly laborious, as most of my readers know, especially when cutting end-way of the grain. For building up segment work the Trimmer has become almost an indispensable tool, and will cut as straight and as clean as it is possible to cut wood. Figs. 1 and 2 are two views of the smallest size Trimmer made by the Fox Machine Co., Grand Rapids, Mich. The illustrations will give the necessary explanation and will be readily understood. The Company make several sizes and the most fastidious "wood butcher" can be suited.

It is not necessary to go into the details of pattern shop requirements, but there is a mechanical paper published that ought to be considered a requirement for pattern makers, and that is the *American Machinist.* Doubtless this is the best and cheapest technical educator we have, for it contains more practical ideas for doing work in all the branches of the machinery business than most papers.

MARKING AND RECORDING PATTERNS.

THE practice of fixing a mark or symbol on a pattern to distinguish it from others, is an excellent one, and the pattern department of any firm cannot disregard it without having much trouble. In many places a large stock is accumulated, regardless of any system, because the man who looks after them calls it "Red Tape" to mark patterns and record them, and boasts of knowing where to find any pattern without any such nonsense, while at the same time they may be piled together like a lot of kindling-wood. What this rule of thumb individual says may be true, about knowing where to find any pattern, but who could find the patterns and know about each piece if he moved off this mortal coil?

The disadvantage that such a firm labors under through not adopting some system of marking and recording is great, and to those who have no system I would recommend the following:—

Fix a raised letter and number on the pattern, so that it shall appear in the casting. The letter is to designate the class of machinery, or it may be used for a certain machine; the number, to distinguish one part from another. The mark that each pattern gets should also be put on the drawings. This is not generally done, but I think if it were, it would greatly facilitate the work in the machine shop. The method which I worked for many years is shown by the sample entry of patterns given on the following pages. The column "pattern at" will be found very useful to those sending out their patterns; it is intended to show where the patterns are. In connection with this column an index is made showing the names of firms with whom business is done. Each firm is given a number, as shown; when a pattern is sent out, the number corresponding to the firm it is sent to is marked with lead pencil in the column, "pattern at," and opposite, the pattern that is sent out; when it is returned the lead pencil mark is rubbed out, showing that it has been returned and stored in its place.

It often happens that in a set of patterns for a certain machine, there are those that will do for other machines; in such cases an entry should be made in the schedule of each machine that this piece will suit, giving the same letter and number of the pattern.

There is always a large number of miscellaneous patterns that cannot be so well classified, yet many of them are often used and need marking; these may be given a symbol and entered under the head of "miscellaneous."

There are also some rough patterns made, the kind that is generally "wanted to be cast to-day." All moulders are acquainted with this kind. It is no use recording such patterns, as they are seldom used the second time; in fact, I think the best way to deal with this class is to break them up.

With large manufacturers carrying patterns for a number of different machines and many classes of machinery, the question may arise what to do when the alphabet is exhausted. When that happens, two letters can be used to designate a machine or a class, commencing with *AB*–

Specimen Entry for Pattern Record Book.

18″ AND 20″ CORLISS ENGINES.

Pattern Mark.		Name of Part.	Castings Wanted.	Weight.	Pattern at	Remarks.
		CAST IRON.				
A.	1	Engine Frame 18″ x 36″	1	3,555	2	This pattern is constructed so that it can easily be changed to different strokes and sizes.
"	1	" " 18″ x 42″	1	3,905		
"	1	" " 18″ x 48″	1			
"	1	" " 20″ x 36″	1			
"	1	" " 20″ x 42″	1			
"	1	" " 20″ x 48″	1			
"	2	Main Pillow Block, - -	1	2,015	3	
"	3	Cap for Block, - - - -	1	475		
"	4	Oil Cover, - - - - -	1	10		
"	5	Corliss Cyl'der 18″ x 36″	1	3,790	2	Patterns are arranged on the telescopic principle.
"	5	" " 18″ x 42″	1	4,290		
"	5	" " 18″ x 48″	1			
"	6	" " 20″ x 36″	1	4,125		
"	6	" " 20″ x 42″	1	4,280		
"	6	" " 20″ x 48″	1			
"	7	Frame Foot, - - - -	1		1	
"	8	Cylinder Feet (20″ cyl.)	2			One Pattern. Loose piece on top for 18″.
"	8	" " (18″ ")	2			
"	9	Crank 36″ stroke, - -	1			
"	10	" 42″ " - -	1			

Specimen Entry for Pattern Record Book.

18″ AND 20″ CORLISS ENGINES.

Pattern Mark.	Name of Part.	Castings Wanted.	Weight.	Pattern at	Remarks.
	BRASS.				
A. 70	Key Heads, - - - -	12			
" 71	Knock-off Latches, -	2			Strong Bronze.
" 72	Conn. Rod Box, - -	1			Crank End.
" 73	" " " - - -	1			Crosshead End.
" 74	Bushing for Piston Rod,	1			
" 75	Bonnet Bushings, - -	4			
" 76	Ecc. Strap Shims, - -	2			
&c.					
	CAST STEEL.				
A. 100	Crosshead Nut, - - -	1		6	Nov. 3d, 1892.
" 101	Cheek Pcs. for P. Block,	2		6	" 3d, 1892.
&c.					

INDEX OF FIRMS TO WHOM PATTERNS ARE SENT.

1	Eddy Foundry Co., ...	Chicago, . .	Iron Foundry.
2	Chicago Foundry Co., ...	" . .	" "
3	Bouton Foundry Co., ...	" . .	" "
4	Webster Mfg. Co., ...	" . .	" "
5	Tarrant & Ramsay, ...	" . .	" "
6	Pittsburg Steel Casting Co., . .	Pittsburg, .	Steel "
7	Solid Steel Casting Co., . . .	Alliance, O.,	" "
8	Eureka Cast Steel Co., . . .	Chester, Pa.,	" "

AC–AD, etc., until through the alphabet again; then begin with *BC–BD–BE*, and so on; thus, it will be seen that it is possible to have a large combination of distinct and separate classes without confusion.

The advantage of recording and marking patterns is that it facilitates ordering the castings and helps to prevent confusion in the foundry. When the order for castings is written out (as it always should be) for the foundry, the mark corresponding to that on the pattern is put on the order so that the moulder and the pattern maker cannot misunderstand each other by naming things differently. Again, when a pattern has a particular mark, every loose piece (and sometimes there are a great many) belonging to the pattern, and also the core-boxes, can be stamped with a mark corresponding to the pattern. The benefit of this is apparent. There is often much trouble caused by not knowing where a certain loose piece belongs, and castings are frequently made minus a piece just because the moulder did not know that it belonged to the pattern; but if every piece is marked as I have said, it leaves no excuse for such omissions. The raised letters that are nailed on the pattern help greatly in checking the castings when received. Especially is the marking of patterns necessary for this, as gentlemen of the quill profession, who generally check the goods, are not usually acquainted with the names of the parts of machinery. Also by this method the finding of patterns is rendered easy even to a stranger; that is if the shelves where patterns are stored are marked with the letter corresponding to the class.

Every firm, large or small, should have some such system as I have described.

PRINTING-PRESS CYLINDERS.

SOME printing-press cylinders have the ends bored out and a short shaft pressed into each end, while others are made with the shaft and cylinder cast in one. Fig. 3 is a section and end view of the latter.

A great deal of trouble is often experienced in getting perfect castings for these cylinders, and to insure good castings they are cast on end in dry sand, or at least they should be; but in spite of all the preventives used against blowholes, dirt, etc., a cylinder will sometimes reveal defects when the first cut is being taken off in the lathe.

Though cast on end, the cylinder is moulded on its side, so that the pattern is made in halves in the ordinary way, as shown in Fig. 4.

As an extra precaution against defects an additional piece five or six inches long is cast on the end, as shown in Fig. 3 from *a* to *b*., the pattern therefore should be made that much longer.

The extra length will receive the impurities of the metal which rise to the top when pouring. It is made thicker on the ends, so that it shall form a head which will exert a pressure, thus helping to produce a clean and sound casting.

Figs. 5, 6, and 7 are three views of the core-box. The end view, Fig. 5, shows it to be built up with staves, which are nailed to three crosspieces, *A*, *B*, *C*. The

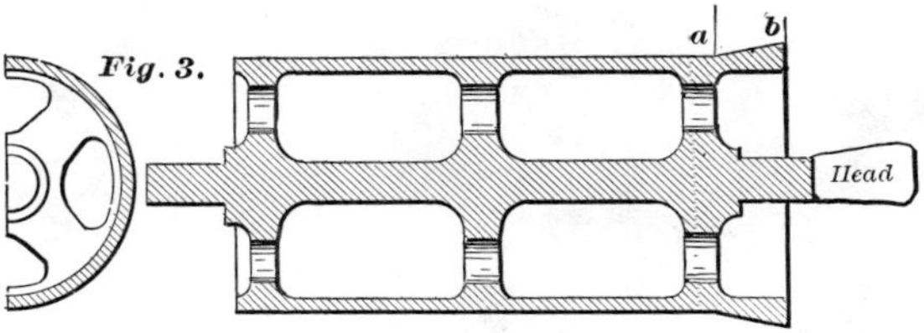

Fig. 3.

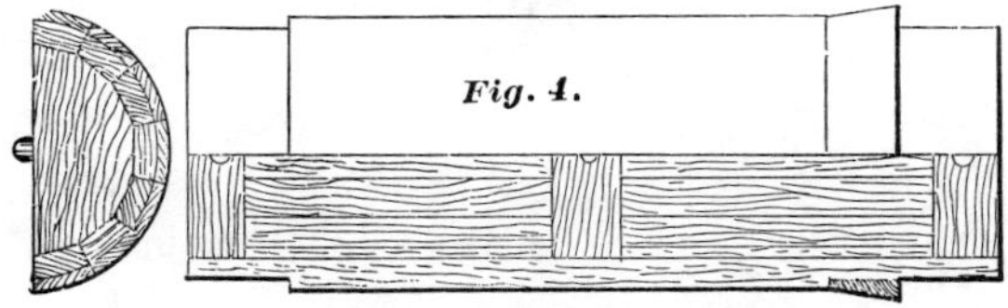

Fig. 4.

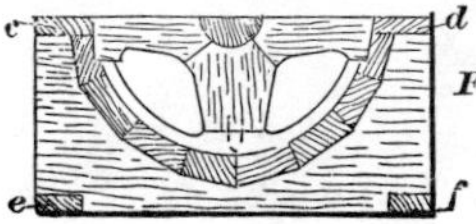

Fig. 5.

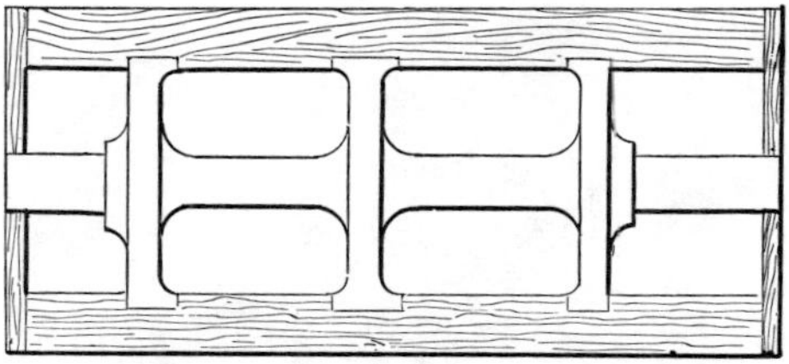

Fig. 6.

Fig. 7.

box is strengthened by running four strips, *c*, *d*, *e*, *f*, lengthwise on top and bottom of the box, fastening them to the crosspieces.

Two of the arms in each set are let in about ⅜″ on the inside to keep them from being rammed out of place, but a dowel pin is put in each of the arms that go in the bottom of box.

This is a very plain and simple job in pattern making and needs no further comment.

DIFFERENTIAL CHAIN PULLEYS.

When the groove in a chain wheel or pulley is made to fit the links of a chain it is sure to be an expensive pattern, especially when made double, like those used in Weston's Differential Pulley Blocks.

Figs. 9 and 10 are two views of one of this kind of pulley. Fig. 10 shows a half section and the pockets for the chain. Fig. 9 is a section through the groove *C, D*, of the large pulley, which has one more pocket than the smaller one. This view also shows how the chain fits into the pockets.

It may not be out of place here to make a few remarks about this celebrated Differential Pulley Block that has so revolutionized the lifting of heavy weights, and for which this kind of pulley was used, in fact, this pulley was the main feature of the patent. T. A. Weston was in this country when he conceived the idea that led up to the Differential Pulley Block.

While Mr. Weston was at Buffalo, witnessing attempts to raise a vessel that had gone down off that city, the thought occurred to him that the necessary power could be obtained from the Chinese windlass, the rope of which winds on two unequal diameters, that is, one half the length of the barrel is larger in diameter than the other. This is practically what this pulley block is developed from.

Fig. 8.

Fig. 9.

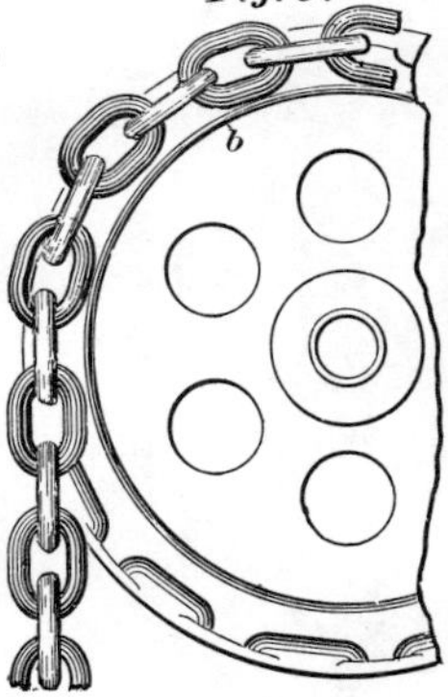

Fig. 10.

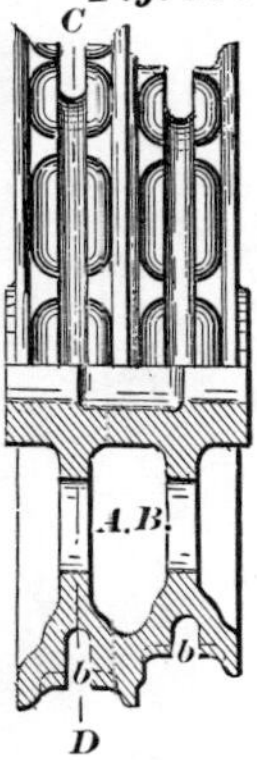

Fig. 11.

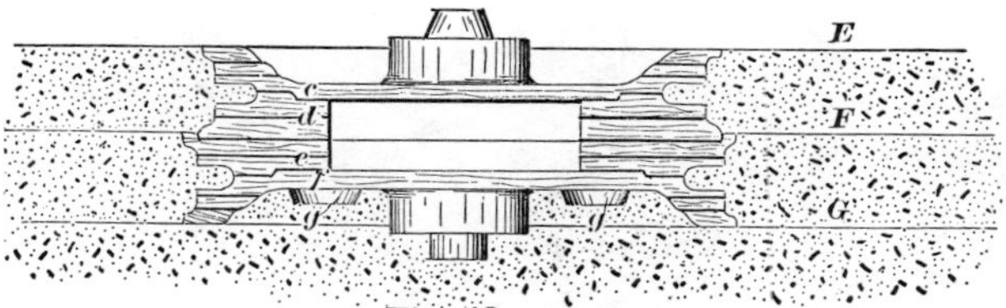

Fig. 12.

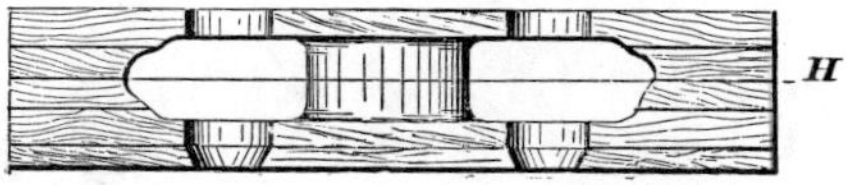

DIFFERENTIAL CHAIN WHEELS.

After much scheming, Weston returned to England and called on numerous engineering establishments, submitting his drawings, but he could find none that would take hold and experiment on his block. Finally he called into a small job shop where the proprietors themselves were working men, paying the extravagant rent of ten shillings a week, and employing six men. That firm has grown since then and employs as many thousands now. I refer to Tangye Bros., Birmingham.

These brothers labored hard to make the block work, and experienced many unexpected difficulties, and when they had perfected it and made it a commercial success, a new difficulty presented itself in the shape of a law-suit in which the Tangyes won. It is pretty hard to conceive of any taller swearing than was practised by the would-be infringers in this case, but I will return to the pattern of the pulley.

They are sometimes made so that all the links of the chain fit into the pockets, but this is an unnecessary expense. The links of the chain that set edgewise in the pulley do not need to fit into pockets like those shown at *a*, in Fig. 8. If the grooves *b b*, in Fig. 10, be turned deep enough to clear these links and pockets made for the other links to set in, it will be sufficient to catch the chain, and will work better than otherwise.

The groove is sometimes formed in a core-box, and a print put on the periphery of the pattern, thus making fewer partings in the patterns as well as the mould; but a much cleaner and better casting can be obtained from a pattern with the groove for the chain cut in it.

Fig. 11 is the section of the pattern and shows how it

is made. The mould is also represented with the cope lifted off, the partings being at *E, F, G.* The pattern is built up with segments and made in four parts, *c, d, e, f.* As will be seen, the casting is cored out at *A. B.*, in Fig. 10.

Fig. 12 is a section of the core-box for this core, and is parted at *H.* The core sets into the round prints *g. g.*; but there are no cope prints, for the reason that it is not easy to close the cope over the six round projecting cores. In the absence of these cope prints the moulder will need to take care that the cope bears on the top of these projecting cores enough to prevent the iron from running in the vent holes of the core, when pouring.

This lightening core can be made in halves or whole, just as the core-maker chooses.

Fig. 13.

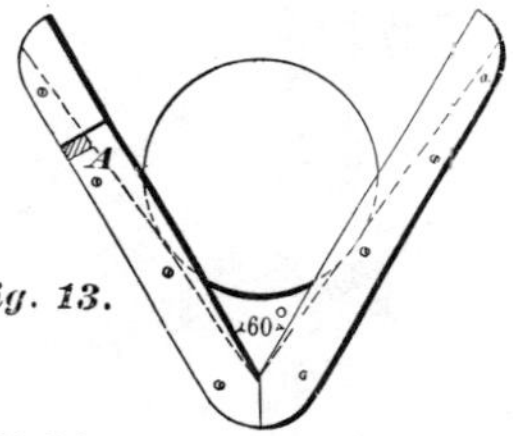

Tool for laying out Hexagon Nuts

Fig. 14.

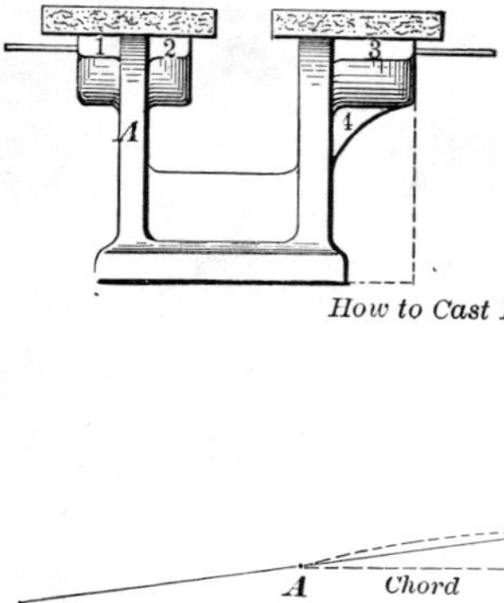

Fig. 15.

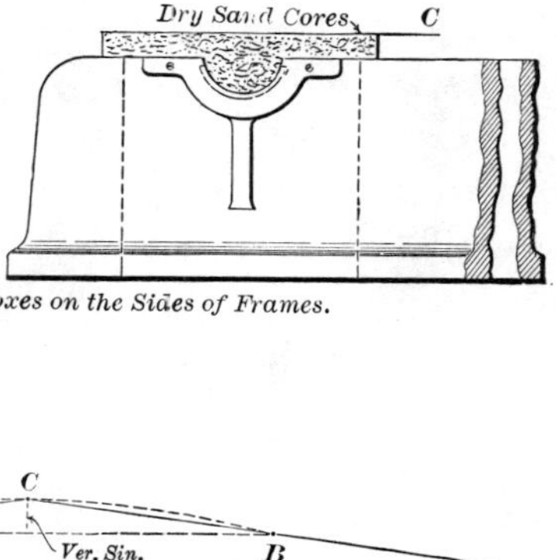

How to Cast Boxes on the Sides of Frames.

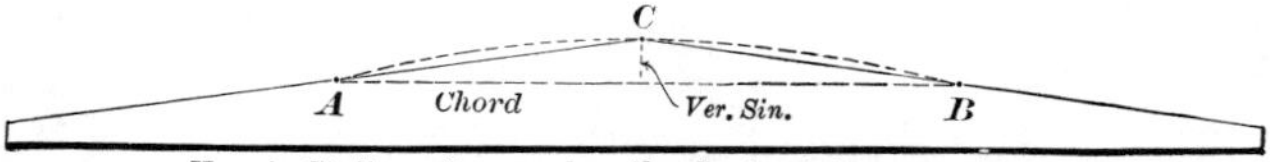

How to Strike a Curve when the Centre is Inaccessible

Fig. 16.

A HANDY TOOL FOR LAYING OUT HEXAGON NUTS.

FIG. 13 illustrates a tool for laying out hexagon nuts, and is very handy to pattern makers; the section of it is shown at *A*. The upper part, which is a light steel blade, is screwed on the lower part, which is made of hard wood and is used in the following manner. After turning pattern to long diameter of nut, place the tool on pattern like a center square, move it round and mark off sides—keeping the two under edges in contact with circle—this is better and quicker than dividing off with compasses and then marking sides.

HOW TO CAST JOURNAL BOXES ON FRAMES.

THE part of a frame shown in Figs. 14 and 15 is one of those jobs that at first looks a little troublesome for moulding, and yet, upon examination the trouble vanishes. The two views show part of a frame with two boxes cast on the sides. The shape of this frame is such as to necessitate casting the boxes down; it will be seen that there is not enough room to draw in the boxes, the sides, *A* and *B*, not being thick enough to allow it. This difficulty may be overcome by making the pattern as shown. The boxes are loose and located on the sides of pattern with loose dowel pins that can be pulled out while ramming up; two cores are made and dried for the boxes, and rammed up with the pattern to *C*, after which the cores are taken out, and the sides of boxes, 1, 2, 3 and the bracket, 4, are drawn. The cores for boxes are then replaced and covered over with sand, the flask rammed up and rolled over. There are other ways of making this pattern; a core print might have been put on the pattern, as shown by dotted lines, and a core-box made with box pattern in it; but the above way of doing it makes a cleaner job. This plan is adopted on many jobs where there is not room enough to draw in loose pieces.

HOW TO STRIKE AN ARC BY THE AID OF THREE POINTS.

IT is sometimes required to lay off an arc, the radius of which is given; but the radius may happen to be so great as to render it inconvenient to locate a center to strike the arc from, or the center may be inaccessible from various reasons; under these conditions the question arises what to do to get the arc.

The following is by no means new, yet it is so little understood by pattern makers generally, that I think it worth while presenting. The all important, in this problem, is to know how to get the point *C*, or the versed sine of arc in Fig. 16; this must be calculated before anything can be done towards striking the arc. Let us suppose the line *A B* to be , say 4 ft., and to represent the chord of an arc whose radius is 20 ft.; it is required to strike this arc without using the center. By using the following formula the desired point can be obtained: $v=R-\sqrt{R^2-c^2}$. This formula need not scare anyone who is not familiar with algebraical expressions; it is very simple, let us examine it. v is the versed sine, R, the radius, c, the semi-chord. v is what we want to get. The formula means that the square of half the chord, which is 4 ft., must be deducted from the square of the radius, which is 400 ft., this will give 396 ft.; then extract the square root ($\sqrt{}$) of 396 ft. which is about

19.9: the formula is now reduced to v=R—19.9. which means that 19.9 must be deducted from R, the radius, 20 ft., this leaves $\frac{1}{10}$ of a foot; $\frac{1}{10}$ of a foot is $1\frac{3}{16}''$+which is the required versed sine.

Having obtained the versed sine of this arc, it is an easy matter now to strike the arc—it is done by cutting a piece of wood to an angle, two sides of which run from point *C* and through *A B*; drive a wire nail at each point of *A* and *B* in the piece on which the arc is to be struck. It will be readily seen now, that, keeping the sides of angle against *A B*, and moving it right and left, the arc can be traced by following with a lead pencil the point *C*. The principle of this is the same as many pattern makers are familiar with—that of using a square for a templet when working out a half-circle core-box.

Fig. 17.

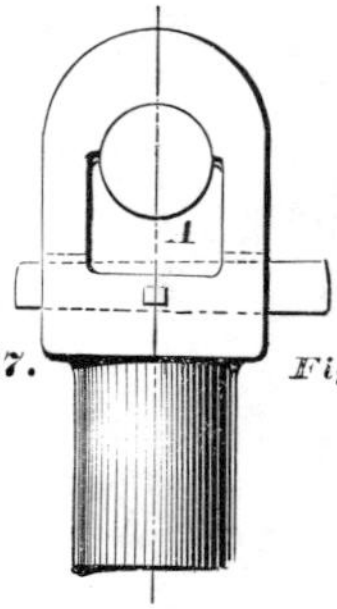

Fig. 18.

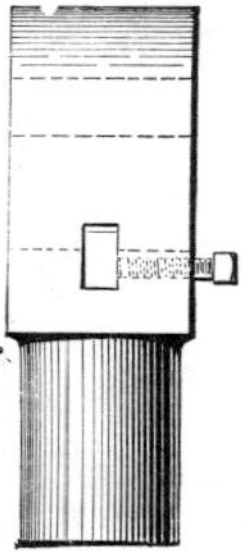

Fig. 19.

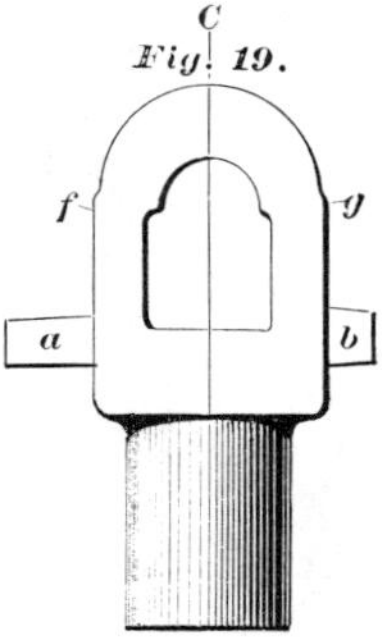

Fig. 20.

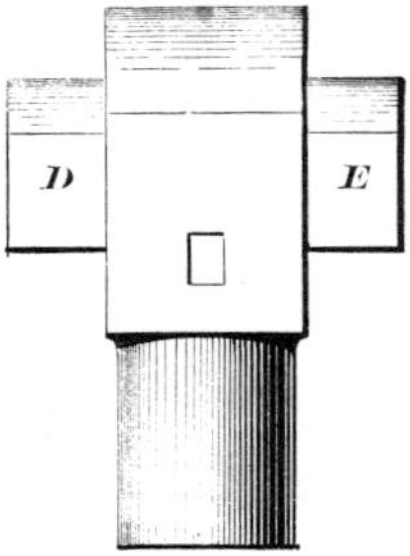

Fig. 21.

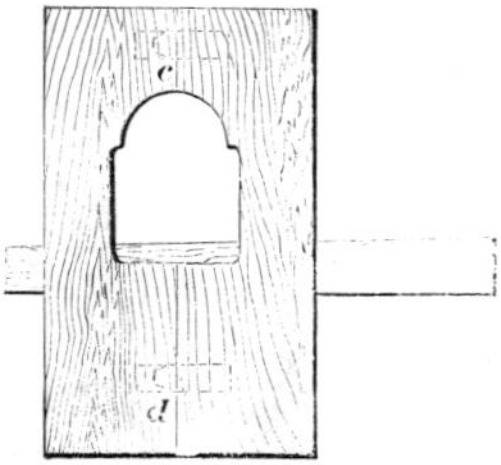

Fig. 22.

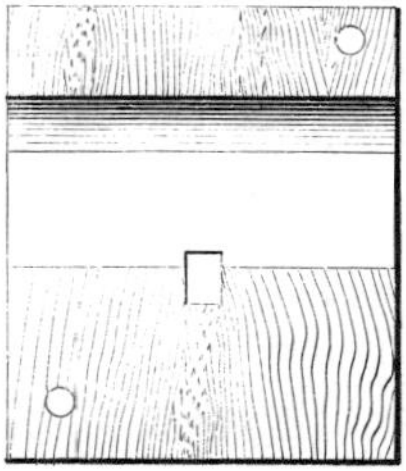

KEY HEADS.

KEY HEADS FOR MOTION RODS.

AN EASY WAY TO LESSEN THE COST OF THEIR PRODUCTION.

THE cost of getting out brass key heads for motion rods may be considerably reduced in the machine shop by the pattern maker doing a little scheming and the brass moulder exercising care in moulding.

Figs. 17 and 18 are two views of a key head, with block *A* in place. A dry piece of cherry or mahogany should be selected and the pattern made as shown in Figs. 19 and 20; it should be in halves, *B C* being the parting line. *D E*, Fig. 20, are core prints which carry the core horizontally. After the core is located a steel key is set in the mould into the print *a*, the cope print *b* bringing the key upright when cope is being closed. The box for the core is shown in Figs. 21 and 22 and is doweled together at *c d*, Fig 21; a key passes through this core-box, which makes a groove in the core to receive the steel key.

It is the intention, when the castings of these heads are being fitted up, to file the round ends, while the sides are to be finished in the machine, therefore the stock allowed for machine finish must stop off at *f g*.

There need be no fear of chilling the sides of the holes through casting steel keys in these heads; the effect is rather the reverse of what would happen in cast iron, for it is well known that if the same thing were to be done in cast iron, the sides of the holes would be so hard that it would be almost impossible to dress them out with a file. A number of keys should be made expressly for casting into the heads, and they ought to be nicely finished, having about $\frac{1}{64}''$ taper sideways; this will enable them to be easily driven out.

I have seen much time wasted in the machine shop making these key heads, such as drilling and chipping out the key hole from the solid, and finishing the round end in a shaper. This led me to devise the above simple way of casting them, which has proved a great saving.

When making large quantities of these castings, a wood pattern will necessarily get broken and badly marked with the moulder's vent wire, so that under such circumstances, a metal pattern and core-box would be more serviceable. I would therefore propose that the standard pattern be made of aluminum and the core-box of cast iron. Aluminum is just the metal that is wanted for such small patterns, because it possesses the two necessary and important elements most desirable for patterns, strength and lightness in weight; a very nice surface can also be made on this metal, which is also the thing needed.

Moulders do not like a heavy pattern, for the reason that it is not so easily drawn as a light one.

Fig. 25.

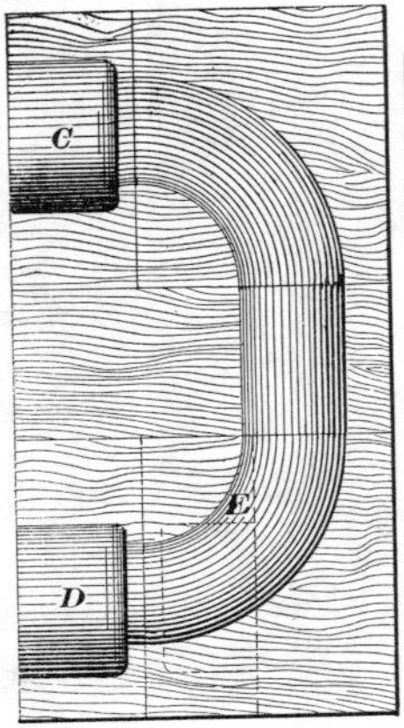

Fig. 23.

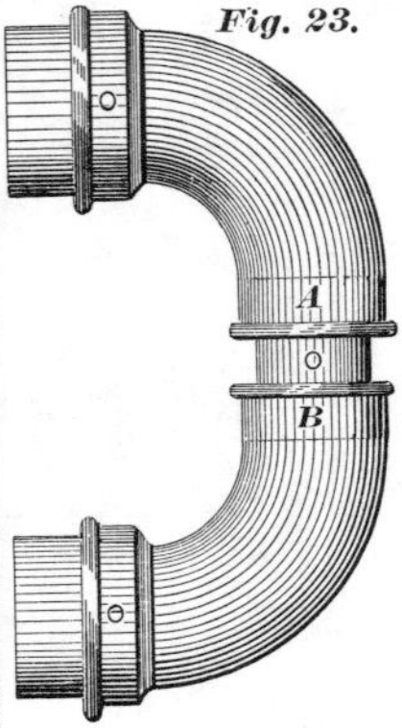

Fig. 24.

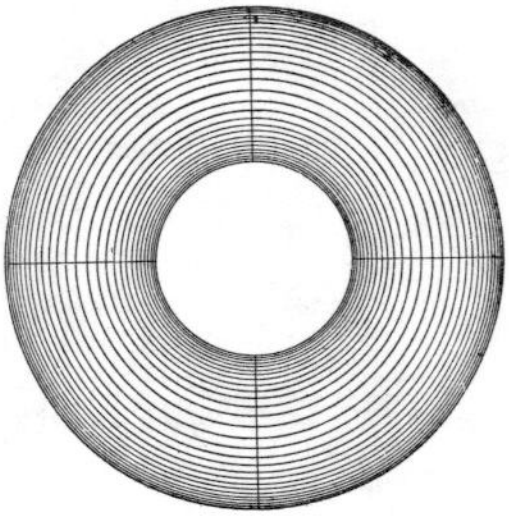

Fig. 26.

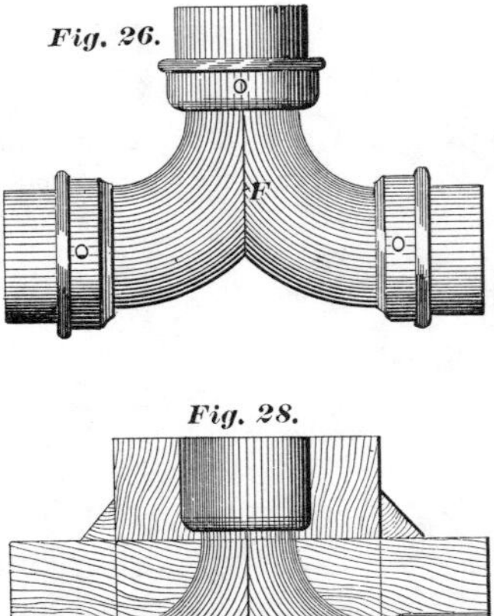

Fig. 27.

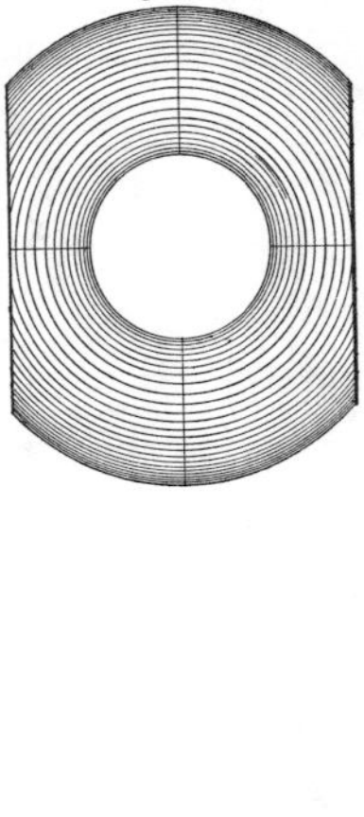

Fig. 28.

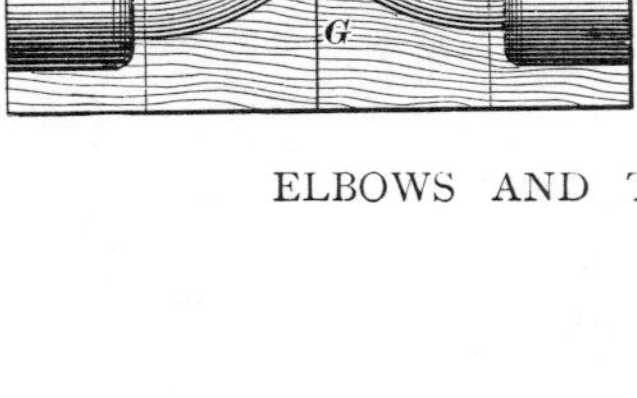

ELBOWS AND TEE PIPES.

ELBOW AND TEE PIPES.

A QUICK METHOD FOR TURNING THE PATTERNS AND CORE-BOXES IN THE LATHE.

MAKING patterns for elbow and tee pipes, if made the right way, is comparatively simple, because nearly all the work can be done in the lathe. For turning out a large number of castings the elbow pattern should be constructed as shown in Fig. 23, so that two elbows can be moulded together. A ring is turned like Fig. 24, the section of it being a half circle, the same size as the pipe; this ring is cut in quarters as shown, and the four pieces used to make quarter turns for the elbows. In Fig. 23, the two spicket ends *A* and *B*, and the sockets, with core prints, are turned on one stick and cut off; the stick should be sawed off long enough to permit of tongues being turned on *A* and *B*, and the sockets, for fastening them to the elbows; dowel pins should be arranged to come one in each socket, and in the print between *A* and *B*, as marked. Fig. 25 represents the core-box, and, like the pattern, most of the work can be done in the lathe and the parts joined together, as shown; a piece screwed on the back will hold the parts together. The core is generally made in halves, so that a full box will not be needed, and therefore only two quarters are used for the turns. In some cases these two quarters that are left may be used to turn the socket ends

C and *D*. The dotted lines show that in this case; the pieces left over cannot be used without cutting out the corner *E* and inserting a piece; in a small elbow it would not be worth while to do this, but for a larger pattern it would probably pay.

Figs. 26, 27, and 28 show the manner of making a pattern for a tee pipe, and will need but little explanation, as it is made much in the same way as the elbow, all the parts being turned and fitted together. In addition to being quartered, the sides of the ring in Fig. 27 are cut off and the parts joined together at *F*, Fig. 26. The parts for the core-box are also cut this way. Before gluing the joint *F* together it should be sized with glue, after which the joint will be very strong when put together; the socket will also help to hold the pattern together at this point. This form of tee is to be preferred to the right angle ones, because of its extra strength and the gradual merging of the passages into each other, which renders the flow of water, etc., easy.

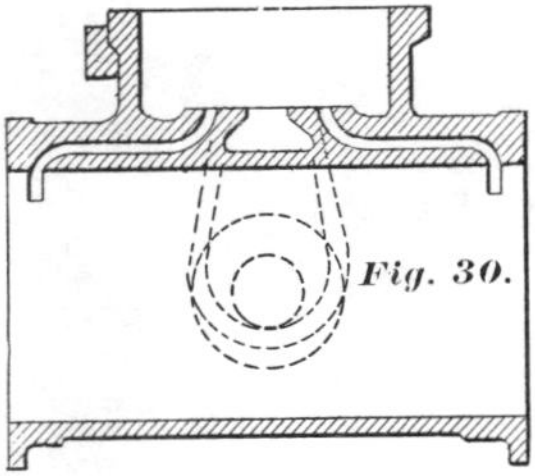

Fig. 30.

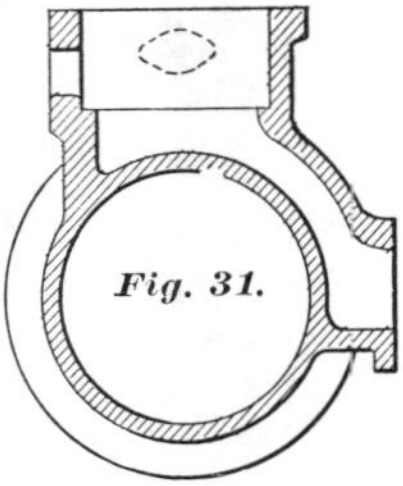

Fig. 31.

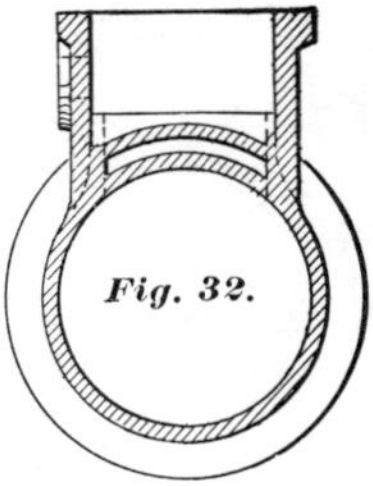

Fig. 32.

Fig. 38.

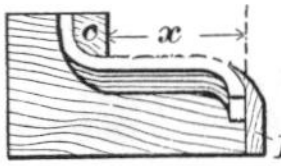

Fig. 39.

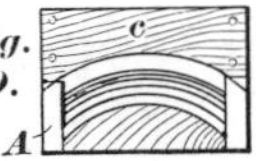

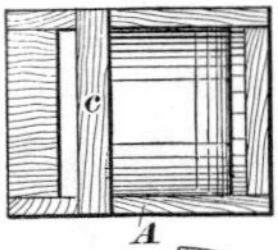

Fig. 40.

Fig. 33.

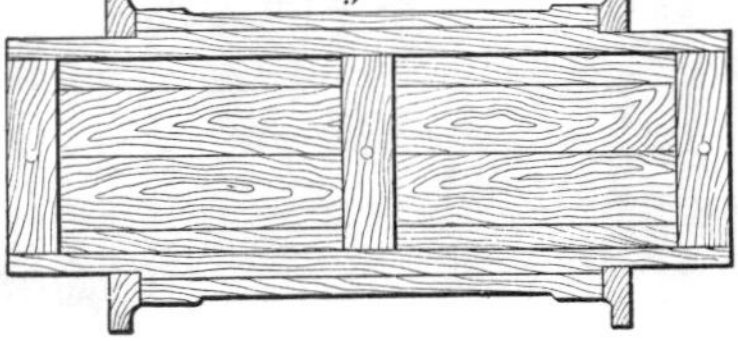

Fig. 34.

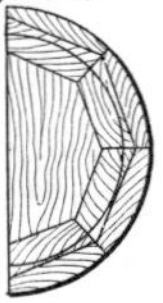

Fig. 35.

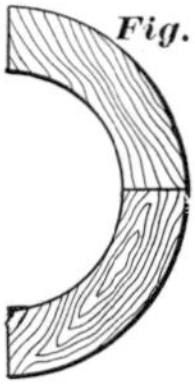

Fig. 36.

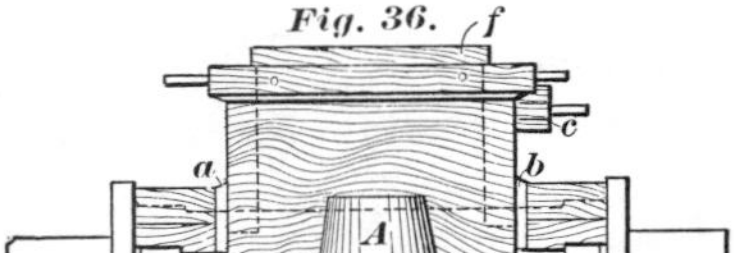

Fig. 37.

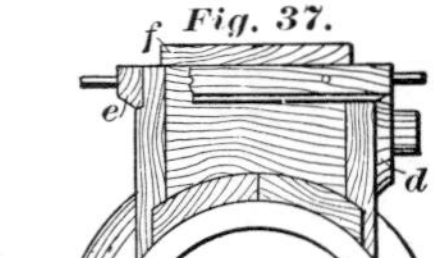

Fig. 43.

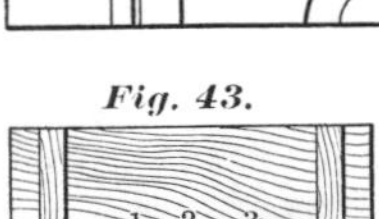

Fig. 44.

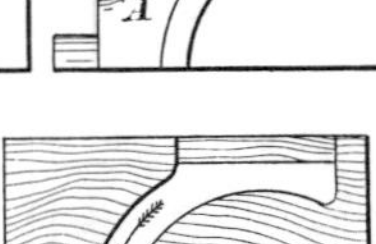

Fig. 41.

Fig. 42.

Fig. 45.

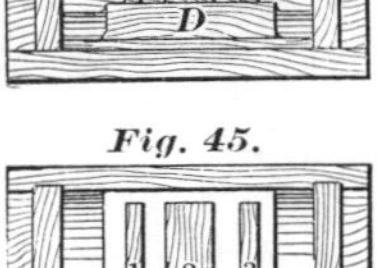

SLIDE VALVE CYLINDERS.

SLIDE VALVE CYLINDERS.

SLIDE valve cylinders are made in a great variety of forms. I have chosen and represented here a well-known type, of which Figs. 30, 31, and 32 are three views. Fig. 31 is a cross section through the steam chest and exhaust port, and Fig. 32 is a section through the steam port.

The way of making the pattern for this cylinder depends largely upon the size of it; if the diameter is to be, say less than 12″, the body of the cylinder may be built up solid, but when above that size it would be better to build the pattern with staves, as shown in Figs. 33 and 34. But one-half of the pattern is shown, which will be all that is needed for explanation. An extra thickness is glued on each stave large enough for turning the body of the cylinder to the required diameter, thus allowing the prints and the cylinder to be in one, which is far better than fastening on the prints. The flanges are made thick enough to turn the fillet on the back of them. They should be got out, as shown in Fig. 35; this will prevent the shrinkage of the wood from affecting the flanges to any great extent.

When the body of the cylinder is built up and turned, the steam chest is made and fitted on, as seen in Figs. 36 and 37. The pieces for the exhaust passage *A*, and those to form a thickness over the ports, are also shown in place. The fillets at *a b*, Fig. 36, are cut out of a thickness that is inserted between the end of the steam chest and port pieces; there is no danger of the fillets coming out when made in this way.

The strips on the steam chest, which give an extra thickness of metal for the studs, are loose and put on with long dowel pins, so that they can be drawn separately, also the valve stem stuffing box, *c*, and the facing, *d*, around the steam opening are loose. The strips are shown at *e*, Fig 37, let in about $\frac{3}{16}''$ around the sides of the steam chest; this is done to prevent them being rammed out of place after the dowel pins have been taken out. The core print *f* should be doweled on, because if it is made fast it is very probable that the moulder will tear it off, for it is a great convenience for him to have this print to place back in the bottom of his mould while dressing it up.

Figs. 38, 39 and 40 represent the core-box for the steam port. Fig. 38 is a side view with the side *A* taken away. The core is swept off on the outside for the length of *x*, but the piece *c* is made to finish the outside, because the core changes from a circular into a a straight part, just where it is entering the steam chest. Fig. 39 is an end view of this box, with the end *B*, in Fig. 38, taken away. Fig. 40 is a plan view and will be understood from Figs. 38 and 39.

The exhaust port core-box is made in halves; one-half is shown in Figs. 41 and 42; the other half is, of course, like this, except that it is made the opposite hand, so that the two halves shall fit when put together. The dotted line in the end view, in Fig. 42, shows how the passage is widened to maintain the same area throughout; that is, it is cut down along the part in the direction of arrow in Fig. 41.

The steam chest core-box is made as seen in Figs. 43, 44, and 45; the side and end are taken off in Figs. 43 and 44, to show the inside; the piece *D* that forms the valve face is screwed on the bottom, and the sides fit over it; this core is the first that is set in the mould, then the exhaust and steam port cores are set into 1, 2, and 3, Fig. 45.

CORLISS CYLINDERS.

SHOWING HOW TO CONSTRUCT PATTERNS AND CORE-BOXES WHICH CAN BE CHANGED AT SHORT NOTICE FOR DIFFERENT STROKE ENGINES.

IT is often required in shops building engines that a cylinder pattern be so constructed that it will serve for engines of different strokes. To illustrate one way of making a pattern like this I have chosen a Corliss Cylinder. Fig. 46 shows the wrist plate side and half section of cylinder. Fig. 47 is a section through *A B*, and Fig. 48 is a section through *A′ B′*. As both halves of pattern are almost alike, it will only be necessary to deal with one half.

Experience has taught me that $\frac{1}{12}''$ to the foot is enough to allow for shrinkage on a job of this kind. Use first quality dry lumber; it will pay to take some pains in selecting stuff for this pattern; for, if the lumber is not thoroughly dry, before the pattern is completed, it will be found that dimensions are scant.

By referring to Fig 49, it will be seen, that to make a pattern for different stroke engines, I have arranged it to slide like a telescope. Fig. 49 is shown pulled apart, and illustrates how the pattern is built. Fig. 50 is a section of Fig. 49 through *C D*; the lagging *a b c d e*, in Fig. 50, is taken off in Fig. 49 to facilitate explanation. An

Section thro' A'.B'.

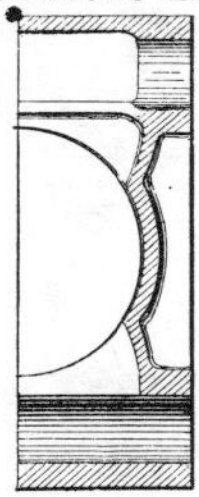

Fig. 48.

Fig. 47.

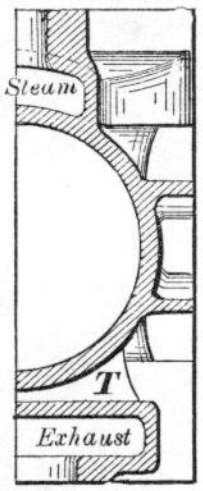

Section thro' A.B.

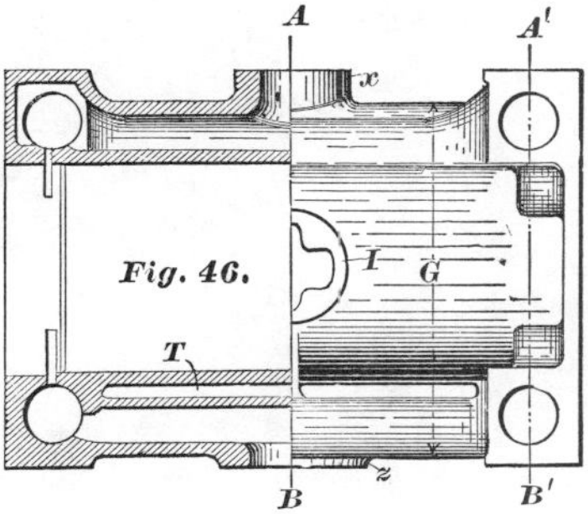

Fig. 46.

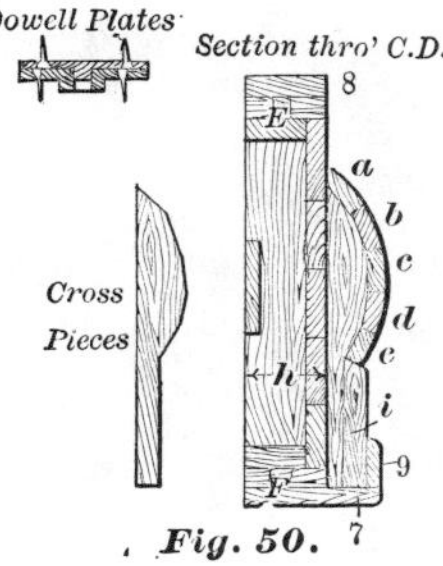

Fig. 50.

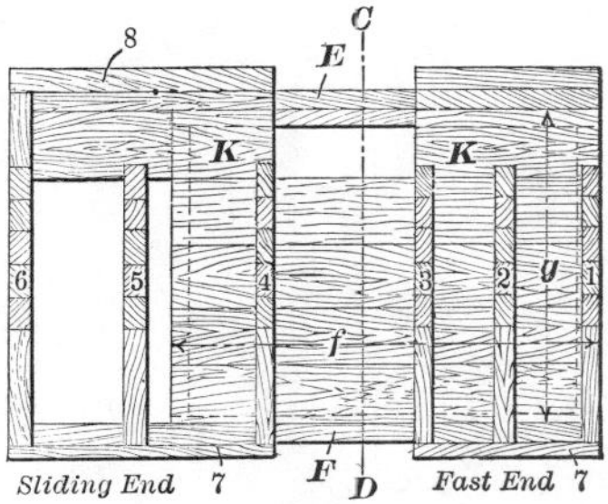

Fig. 49.

Fig. 53.

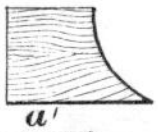

Fig. 54.

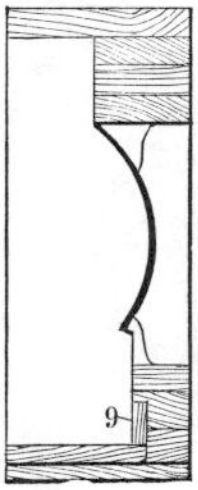

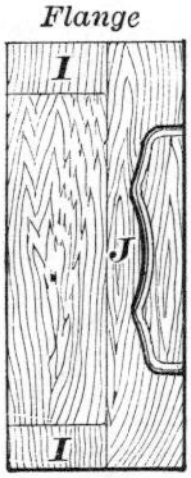

Fig. 52.

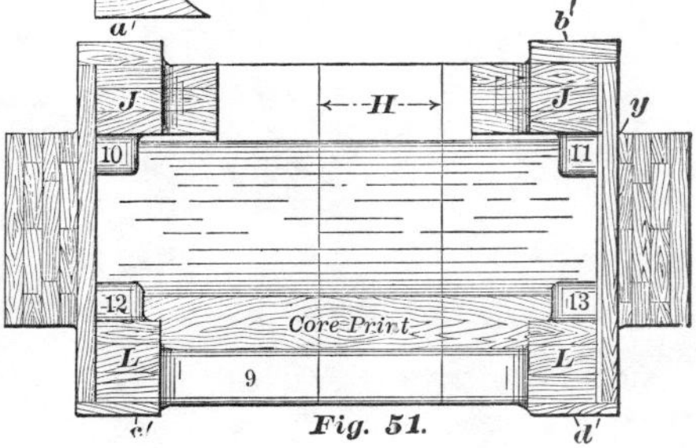

Fig. 51.

CORLISS CYLINDERS.

inner box is first made, the length, width, and depth of *f g h* in Figs. 49 and 50, and doweled together with iron dowel plates like sketch. Build out each side of this inner box for about two-fifths of its length to width *G*, in Fig. 46, then fasten on strongly three cross pieces numbered 1, 2, 3. For future reference I will name this the "Fast End."

In Figs. 49 and 50, the outside piece marked 7 should be made wide enough for the exhaust passage in Fig. 47. Notice that in making the inner core-box the board *K* is cut in two so that one piece may form a part of the sliding end. On the sides of box, glue and screw two guides marked *E F* in Figs. 49 and 50; on each side of these guides, fit pieces that shall slide over them, and secure them to the two outside pieces, 7 and 8. This will make the sliding end the same width as fast end. Now fasten the cross pieces, 4, 5, 6, and piece *K*, to the two slides, and be careful not to get any of them glued to the box; the lagging *a b c d e*, in Fig. 50 will hold this sliding end together, and, as the pattern progresses, other parts will make it secure. The piece *i* forms the core print for coring out the space *T*, that separates exhaust passage from body of cylinder. (See Figs. 46 and 47).

In Fig. 49 we have now a foundation to build on, and have made it the width of *G* in Fig. 46. At this stage, the length of the pattern, when closed for the shortest cylinder required, should be the length of the cylinder, minus the thickness of two flanges and $1\frac{1}{4}''$ for fillets on back of flanges. Fig. 51 represents the pattern closed, with a filling piece, *H*, put into it and the framework of Fig. 49 all closed up.

Proceed next to get out the end flanges, valve chambers, port pieces, etc., and build on. Get out the flanges, as shown in Fig. 52, glue on three pieces, *I, I, J*, $\frac{5}{8}''$ thick; this $\frac{5}{8}''$ thickness is for carving the fillets on backs of flanges, and makes by far the best job. It must be remembered that the cylinder illustrated here is supposed to be a standard pattern, from which a large number of castings may be taken, so that in this case the best way will be the cheapest in the end.

Glue and screw on the end flanges, shown in Fig. 52, and the side pieces, a' b' c' d', Fig. 51, for the valve chambers. Build some blocks together, same shape as shown in Fig. 53, and fasten them in place at *J J*, to form the ends for steam valve chambers. For very large cylinders these pieces had better be boxed. For the ends of exhaust valve chambers, *L L*, make plain square blocks; now make the exhaust passages the right height, by gluing on piece 9 in Figs. 51 and 54; after this, fit in the pieces 10, 11, 12, 13, Fig. 51, that give a thickness of metal over the steam and exhaust ports. All is now ready for rounding off the corners and side of steam passage, and cutting the fillets.

Check down the two inner edges on ends of valve chambers (See Fig. 46). This is to allow the bonnets to lap over and cover the joint of the black walnut lagging, that these cylinders are generally cased in. I have said that both halves of pattern were almost alike; the difference between them is, the center piece for bolting the wrist plate to is usually on the opposite side of the small bosses that take the indicator pipe; also the valve chambers are made about one inch longer on the cope side of the pattern than on the drag side; this is done to insure solid ends after this extra inch is planed off the casting.

When casting these cylinders, all dirt and flux that is in the mould and metal rise to those four high places and stick there; hence, the necessity of extra stock on the ends of valve chambers.

In turning the eight round core prints for the valve chambers make those in the cope as much shorter than those in the drag, as the extra stock just mentioned; in other words, the core prints should measure the same length from the joint of pattern on both halves. The necessity of this will be seen when we come to make the core-boxes. Give these prints plenty of taper, because it will help the moulder to feel his way when dropping in the cores of valve chambers; it will also help to bring the cores upright when dropping on the cope.

Now fit on the pieces *x* and *z*, to take the exhaust and steam flanges (see Fig. 46), and, as these pieces are to be removed for sliding the pattern out and in, they must only be doweled and screwed on. Screw the core prints on these pieces from the back. The center piece, *I*, Fig. 46, should only be doweled on. Build up the core prints for bore of cylinder, as shown, and turn a flat fillet on them at *y*, Fig. 51. This will prevent crushing the edge of sand when putting the core in and dropping on the cope. Turn these prints whole and saw in halves with band saw after turning them; for large cylinders it will be better to build these prints with staves. We may now consider the pattern finished.

There are many details which are omitted, because

they are such as the pattern maker will naturally run against; so let us leave the pattern and start in on the core-boxes for the steam and exhaust passages, valve chambers, etc.

It will scarcely be necessary to enter into the minor details of making the core-boxes, seeing that I have shown both cores and core-boxes for the steam and exhaust sides of a cylinder.

However, a few explanations may be necessary. Like the cylinder pattern, I have only shown half sections of cores; Figs. 55, 56, 57, are three views of the core for steam passage and valve chambers. A half core-box is all that is needed to make this core, the joint of it being on line *C′ D′*, and the joint of core-box on line *E′ F′*.

Figs. 58, 59, and 60 are three views of the core for exhaust passage and valve chambers, also made with a half core-box.

The core-box for the steam core is constructed as shown in Figs. 61 and 62. The port cores, *e′ f′*, in Fig. 57, are made separate, and pasted in; the core prints to receive these port cores are seen at *e′ f′* in Fig. 62. Like the pattern, these boxes are made so that the length can be changed. In Fig. 62 the joint, *g′*, for the length of *h′*, is not glued, there being a tongue or guide for this surface to slide in; the piece, *i*, is loose, which can be taken out, thus allowing the valve chamber part to be slipped toward the center as may be required; on the inlet side of this box, the joint *m*, for the length of *o*, and the joint *n*, for the length of *p*, is also made to slide, and when the box is to be shortened, the two pieces, *r* and *s*, are taken out, which will allow the valve cham-

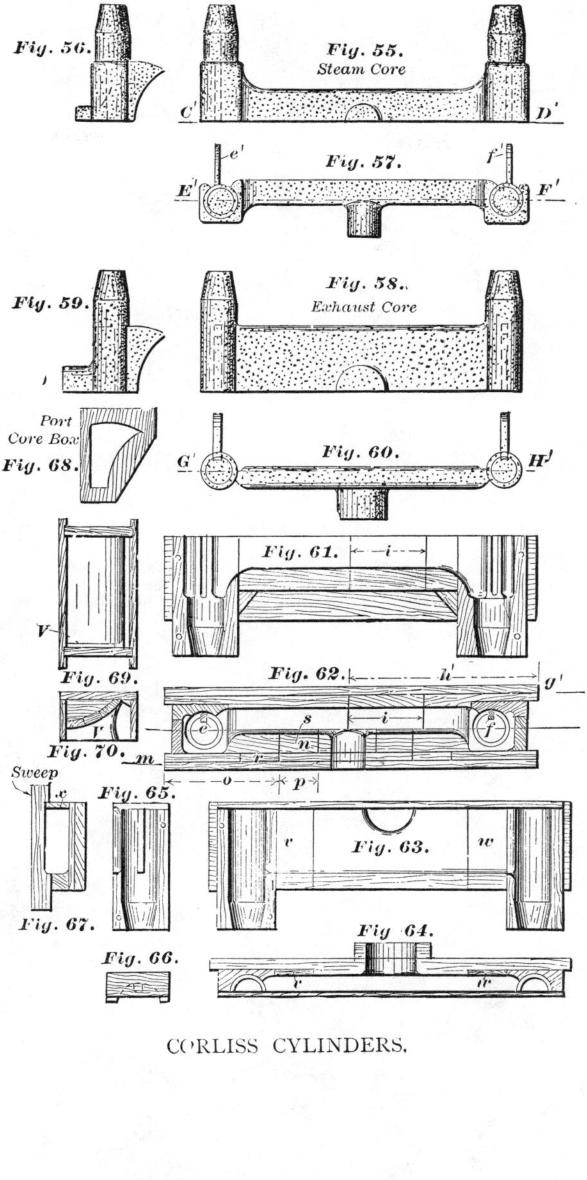

CORLISS CYLINDERS.

bers to be slipped toward the center; each side of the inlet is alike. The reason there are more loose pieces on this side of the core-box than the other, is because the inlet must always be located midway between the two valve chambers.

The exhaust core-box of which Figs. 63, 64, 65, 66 and 67 are five views, is made open on one side between the valve chambers, and as this open side is a plain straight surface, it can be swept off with a sweep, as shown in Fig. 67. The valve chambers are round, all the way through, in this core, as will be seen by referring to Fig. 58. Like the core-box for the steam side, two pieces, *v*, *w*, Figs. 63, 64, are inserted, for changing the length. Fig. 67 shows a section of the box; the side *x*, is carried above the center line of valve chambers.

Fig. 60 shows the side of core runs over the center line *G*, *H*, hence the necessity of carrying one side of the box above the center of valve chambers.

Figs. 65, 66, are two views of the port halves of valve chambers, and, like the steam core, the port cores are made separate and pasted in.

The core-boxes for the exhaust and steam ports are made like Fig. 68, except that the exhaust port box is made thicker than the steam port box. In small size Corliss Cylinders, whole core-boxes are sometimes made, instead of half-core boxes, but for large cylinders, the cores are better made in halves both for convenience of making and drying.

The box for coring out the space that separates the body of cylinder from the exhaust passage may be made as shown in Figs. 69 and 70. Make the box half the length of the space. (See Figs. 46 and 47 at *T*). It will be seen that in the center of this space, there is a bridge; to form this, a loose piece, *V*, Fig. 69, half the thickness of the bridge, is made and fitted in the end of core-box, and as there must be two right and two left hand cores, changing this loose piece to the opposite end will give it. It will be found that two other small cores are needed, one right-hand and one left-hand, to core out between the valve chambers and cylinder on the exhaust side. These small cores are really a part of the core for coring out the space *T*, in Figs. 46 and 47, and might be made in one core, like the core print shown in Fig. 51, but I think it is easier to make them separately. I have not shown these small boxes, as they will be readily understood without.

Those who make Corliss Cylinders, will no doubt see the advantage of making their patterns to slide the way I have shown, as the pattern can be easily and quickly changed for different lengths without damaging it.

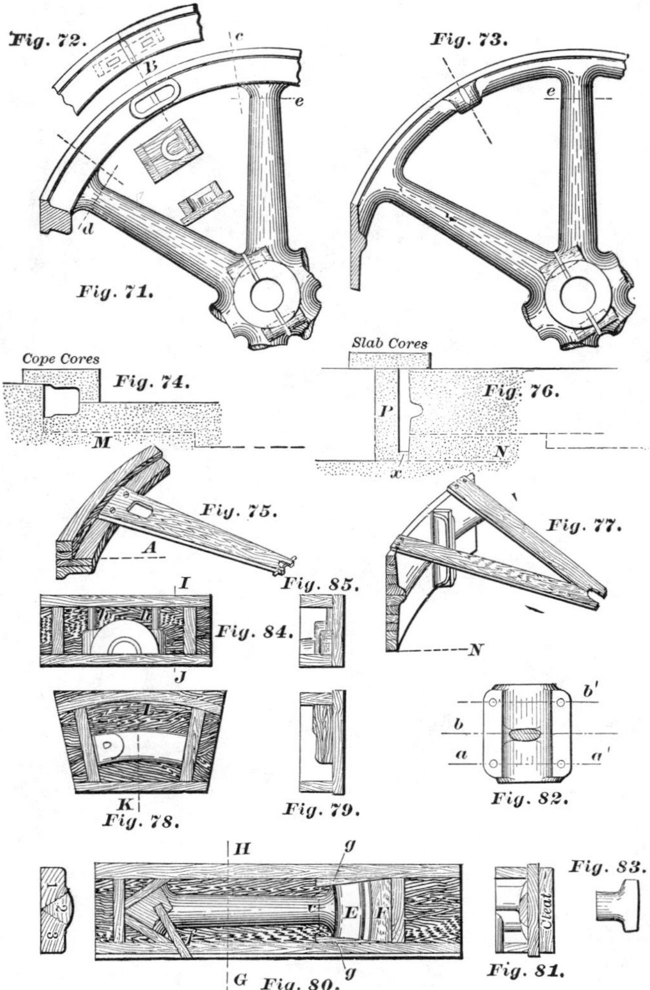
Fig. 72.
B
c
e
d
Fig. 71.
Fig. 73.
e
Cope Cores
Fig. 74.
M
Slab Cores
Fig. 76.
P
N
x
Fig. 75.
A
Fig. 77.
N
Fig. 85.
I
Fig. 84.
J
K
Fig. 78.
Fig. 79.
b'
b
a
a'
Fig. 82.
H
g
E
F
G
Fig. 80.
1
2
3
Cleat
Fig. 81.
Fig. 83.

FLY WHEELS.

FLY WHEELS.

DIFFERENT STYLES.

MAKING fly wheels is such an every-day occurrence, that it seems almost unnecessary to say anything about it, and yet I think something might be said that might be of service to some one.

These wheels are generally cast in halves and bolted together, except in the case of a fly wheel with square rim, in which case it is generally held together with wrought iron links, shrunk on each side of the rim, as shown at *B*, in Fig. 71, or sometimes with cotter-bolts, as seen in Fig. 72.

Fig. 71 is part of a square rim fly wheel, and Fig. 73, that of a band wheel. These two wheels are the same diameter and of about the same weight. Let it be supposed each is designed for one size engine; one customer will want a band wheel for his engine and another may want a fly wheel with square rim, and because of this, it is proposed to make the same arm core-box, with changes, do for both wheels. Fig. 74 is a section of a mould for a square rim wheel; it will be seen that the lower and outside part of the mould are formed in green sand, the segment pattern in Fig. 75 being used for that purpose. A step is made in this segment at *A*, so that in ramming up, the green sand can be swept off level with it; this makes a level surface on the inside, on which to set the cope cores shown in Fig. 74. A pin is put through the end that fits over the spindle to prevent it from getting away.

Fig. 76 is a section of the mould for the band wheel; the inside of the rim is all rammed up in green sand, using segment in Fig. 77. Two views of the core box for the cope of square rim fly wheels are shown in Figs. 78 and 79, and the length of the box is from *B* to *C*, in Fig. 71. *D*, in Fig. 78, is a print to receive the core for the link, and as there are to be two right and two left hand cores, it must be changed to the other end of the box for opposite hand cores. Two views of the core-box to form the recesses for links are shown between the arms of Fig. 71, and cores made in this box will be set in print *D* and also in a print on the segment.

The arm core-box, of which Figs. 80 and 81 are two views, should be made very strong, as it, of necessity, gets very rough usage. If this arm box is built the way I have shown, and a bolt put at each end to hold the sides together, it will stand a considerable amount of rough handling before coming apart. The depth of the box is governed by the distance of the bolt holes in the hub from the center of arm; in this case *a*, *b*, in Fig. 82, is the depth; notice also the box is made longer than it is needed for these wheels. This is done that it may be used for larger wheels. *F*, in Fig. 80, is made loose and is laid on top of *E*, which is also loose; the piece, *F*, is used in the box for the lower half of arm cores, and for the upper half it is taken out, because the ends of the upper arm cores must lap over the outer edge of mould in the same manner as the cope cores for the rim. (See Fig. 74).

For the band wheel, *E* and *F* are taken out and the piece in Fig. 83 substituted. The end of the box is then made to fit the inside of the rim for band wheel. The length from center of hub to *C*, in Fig. 73, corresponds with the distance from the center to dotted lines, *d*, *e*, in Fig. 71. The three pieces marked 1, 2, 3, are used to form the hub in arm core-box. In Fig. 80, *f* is a half round core print and is changed to the opposite side when piece numbered 3 is used, and on a wheel having six arms there will be twelve cores, four of each from the pieces 1, 2, 3. In order that the cope cores for the rim may fit against the sides of arm cores, two wedge pieces, *g*, *g*, are fitted against the sides of the box, making that part of the box radial, like the ends of the box in Fig. 78 are made. In making the core-boxes, a clearance, (say $\frac{1}{16}''$ on a side) should be allowed where the cores fit together at the hub and rim; nine times out of ten the moulder has to file these cores to get them in position, and it is very provoking to find, when the last arm core is being set, that it will not go in place by $\frac{1}{2}''$ or more.

The box, in Figs. 84 and 85, is to complete the outer part of the hub on each side of line *a′ b′* in Fig. 82. *h h* are half round prints that will match *f*, in Fig. 80, when cores are set.

In Fig. 74, the dotted line, *M*, represents the surface of a level bed that is struck, and which is the first thing to be done towards making the mould, but before the rim can be made, the core for the end of hub must be located, the top of it being set flush with the bed, *M*. This core, which forms part of the hub, will be a guide to set the lower halves of arm cores, which is the next thing to be done. After this, the green sand part of the mould is rammed up.

The band wheel is made in much the same manner, except that in Fig. 76, the level bed is struck off at *N* and the arm cores blocked up to the proper height, and the inside of the rim rammed up.

The outside of the band wheel can be formed by cores, like *P*, being set around. When making the segment, it should be made deeper from the center rib to *N*, than it is to the cope edge. The reason for this is to allow for the piece *x*, that is made on the core, *P*, *x* being a guide for setting these outer cores, so as to give the desired thickness to rim.

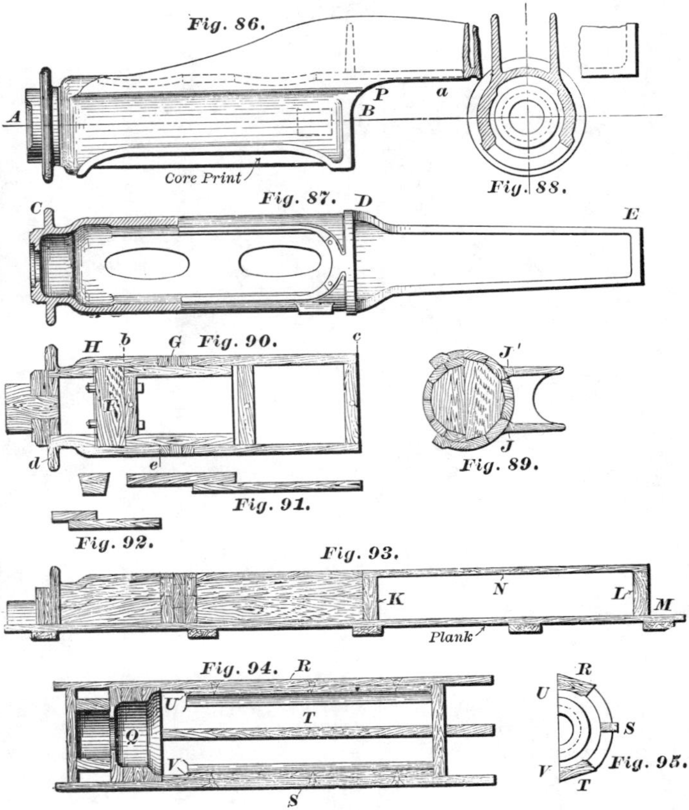

ENGINE FRAMES.

ENGINE FRAMES.

HOW TO BUILD THE PATTERN TO SERVE FOR VARIOUS STROKES.

The type of frame shown in Figs. 86 and 87 is the same as used on the Corliss Engines built by the M. C. Bullock M'f'g. Co. Chicago, and is considered by mechanical engineers to be a strong frame. Among a number of ways, it may be difficult to decide which is the best way to construct the pattern for both pattern maker and moulder, for in all such jobs the pattern making and moulding should be considered together.

Figs. 86, 87 and 88 are three views of this frame to be built. Fig. 87 is shown in part section towards the cylinder end of frame and Fig. 88 is a cross-section showing the two ribs on the back. The pattern is to be parted on line *AB*, Fig. 86, and to mould with the two ribs down. That part from *AB* to *a*, can be lifted with the cope, or it can be lifted out with an anchor plate and set in on chaplets. This pattern is also made on the same principle as the Corliss Cylinder, viz., to slide. By doing this, different stroke engines can be made with the same pattern. Provision for this change must be made between the points *CD* and *DE*, in Fig. 87, but the part between *CD* will only be made to slide, while the length *DE* can be changed by making a false end to be used for stopping off to the length required.

We start the pattern by building up two half round pieces, one to be larger in diameter than the other; the smaller will be the diameter of the inside of frame, and will therefore serve as a core print, and the larger for the outer diameter of frame. The end view of these two parts is shown in Fig. 89, their length is from *b* to *c*, in Fig. 90. Notice that on one end of the larger part, it is turned down to the same diameter as the small part; this is for the sliding end to slide on.

Get the staves out for building the larger part, and glue on an extra thickness, forming a step as shown in Fig. 91. After building the two parts, then put them together and turn, but as one part is larger than the other, they must be balanced before turning. It will be seen that only the large piece and the end of the smaller can be turned; the remainder of the smaller piece will have to be planed off. Next get some staves out like Fig. 92 to make the sliding end *H*, as shown in Fig. 90. The length of this will be from *d* to *e*. The flange is glued up in two courses and fastened around this sliding end, *H*. The end piece that receives the cylinder should not be fastened on permanently, as one frame does for two or more sizes of cylinders, so that instead of putting on and taking off a lining, it is better to make two or or three ends for changing.

In Fig. 90 the pattern is shown arranged for the longest stroke. *F* is a piece bolted on the end at *b* to give more bearing to the sliding end, *H*. *G* is a filling piece built up in three courses so that whenever it is necessary to shorten the frame, these two pieces, *F* and *G*, can be easily taken out, and the sliding end moved up and screwed again. By referring to Fig. 86 it will be seen that the half of this sliding end that goes around the print has to be cut out circular, like the front end in Fig. 87, and moulding fitted around it. In making the moulding that is shown around the edge of frame it should be made loose from the print as far as *o*, *o*, thus making strong loose pieces, which are less likely to be broken.

That part of the frame that covers the print should now be fitted on. Having the round part of frame ready, get the ribs built on the back by first planing two flat places on the pattern where these ribs connect to the body, wide enough to take a piece that shall form the inner and outer fillets at bottom of ribs; see *J*, Fig. 89.

Fig. 93 shows the large part of pattern; it is laid on a plank that is surfaced; on this plank a center line is struck as a guide for getting the other part of frame in line with the body; after locating the center line on pattern to that on the plank, fix a temporary piece, *K*, on the end, and also a piece, *L*, the same height. The point, *M*, shows the full length of frame; these two pieces, *K* and *L*, are temporary supports for locating the straight part of frame. On these supports square up a center line from the plank. The straight part, *N*, should be flush with the two pieces, *J' J*, in Fig. 89, so that the two ribs can have a straight bearing from end to end.

Now is the best time to bore some holes through the body of pattern for screwing on the ribs. It will strengthen the pattern to glue some dowel pins through the body and down into the ribs. Having put this part together securely, proceed to fit in the bracket at *P*, Fig. 86, remembering that the inside of these brackets should match the inside of frame. The two lightening holes at the back, seen in Fig. 87, are cored. These cores will serve to set the main core on.

Having completed the pattern, make a half skeleton core-box for the inside of frame. This box is seen in Fig. 94, of which Fig. 95 is a cross-section. First build an end as shown at *Q*, and though but one-half be needed, it will pay to build a full circle and turn it. After cutting it in halves, take one and screw on three pieces, *R*, *S*, *T*. To *R* and *S* screw on two other pieces, *U* and *V*, for the guides. When the core is being made, the screws shown are taken out, thus leaving *U* and *V* behind to be drawn out separately. This skeleton box is arranged for the longest, and when changing length of frame all the change that the core-box will require is to fit a piece in the plain end to the desired length.

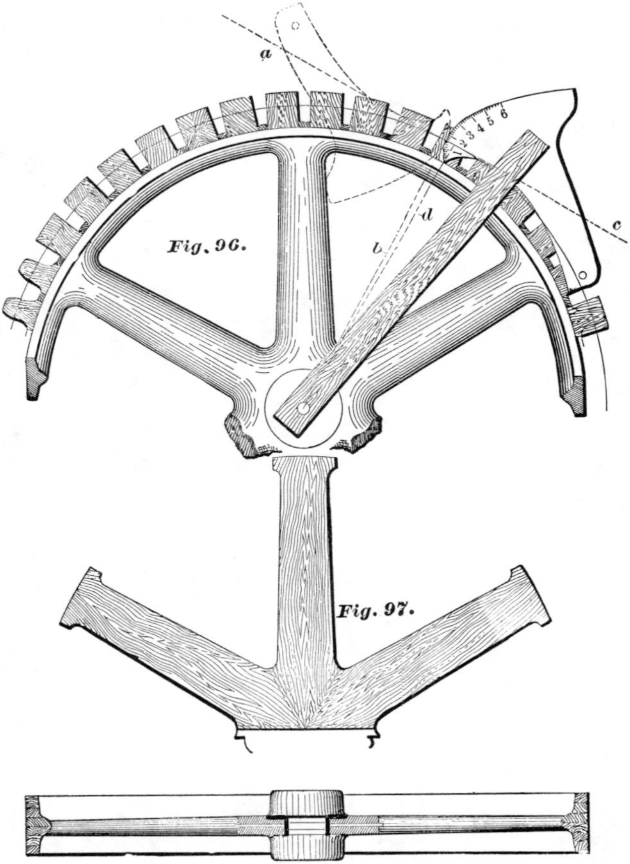

Fig. 98.

SPUR GEARS.

SPUR GEARS.

AND HOW THE TEETH SHOULD BE MADE.

There are so many ways of making gear patterns that it would be extraordinary to show some new way. The question is, which is the best way to make a gear pattern that will stand a reasonable amount of hammering, not be unnecessarily expensive, and so that a casting when taken from it will run smoothly. Even on a gear pattern much unnecessary time may be spent, and is spent, that does not make the pattern any better, but only more expensive. For instance, dovetailing the teeth on gears $1\frac{1}{2}''$ pitch and above, is unnecessary work. Much time can be saved by fastening the teeth on, as shown in Fig. 96. I have found this way to be cheaper and by no means inferior to dovetailing.

It is often found that after the rim has been turned and the teeth dovetailed on, that the wheel runs out; there is no doubt that the cutting of the rim and driving in the dovetails have to do with this, and yet there are men who will argue that a gear cannot be made right unless the teeth are dovetailed on.

By the method shown in Fig. 96, we have the advantage of getting a fillet at the root of the tooth, which cannot be well made on one that is dovetailed on; in this way the tooth is also strengthened and made better for moulding, which means a better casting.

In making the pattern, the blocks for the teeth should

be sawed out first, and then the segments for building the rim. These should be laid aside for awhile to allow any moisture that is in the wood to dry out, and though the stock may be considered ever so dry, it is better to do this, and the importance of it cannot be over-estimated in gear making.

In the meantime get out the arms and put them together; if the arms are of an oval section, now is the time to shape them, because it is far easier to do this before fixing them in the wheel.

Fig. 97 represents the way the arms may be put together. It will be seen that tongues are inserted in the joints where the arms come together. The grain of these tongues should not run parallel with these joints, but across.

Before building on the last two courses of segments for the rim, turn the rib that is on the inside (see Fig. 98), then build the arms in place, taking care not to fit them so tightly as to spring the rim—the remaining courses may now be laid up.

In Fig. 98 I have shown a cast iron flanged bushing; the center of arms is turned out to receive it, and is secured by one or more screws in each arm. I would suggest here that it would be a good thing to have a bushing of this kind in nearly all wheels, and that a standard sized hole be adopted. All hubs should then have a projection turned on them to fit the standard size. There would be but little trouble then in changing the hubs.

After turning the rim for the wheel, fit on the blocks for teeth, and screw them on from the inside of rim; then fit strips between, as shown; take care not to allow any glue to get between the blocks and strips when nailing and gluing on the latter, because the blocks have to be taken off again to work the teeth out.

In some cases it may be better to screw on a block and fit a strip the right width against the side of block before screwing on the next one, but this is purely a matter of choice with the pattern maker.

There is quite a variety of systems employed for laying out the profile of gear teeth, and I do not propose to enter into any discussion regarding the merits of these different systems; but much of the controversy is so hair-splitting that in practice it amounts to nothing when applied to cast gears. In some shops the teeth are laid out to "suit the eye." Of course this is entirely wrong, and, though such gears may run, there will be much jarring and friction, which means extra wear and tear, and loss of power.

For general purposes it is desirable to get that form of tooth that shall work with the least possible friction, and at the same time be interchangeable with any other of the same pitch. There are various ways of getting this. One good way is by the use of Prof. Robinson's Templet Odontograph. Fig. 96 shows how to apply the Odontograph. The full lines show it in position for marking the face of tooth, while the dotted lines show it reversed for marking the root.

To locate the Odontograph in proper position for the face of tooth, "setting" tables are supplied with the instrument, which will give the graduation on the Odontograph to set to pitch line; but this graduation is not all that is needed to set the instrument by. In Fig. 96, two dotted lines, *c* and *d*, are drawn from the *center of the tooth*, forming a right angle, *d* being radial; now, by setting the hollow edge of the Odontograph even with the line *c*, and using the graduation referred to. the location is determined for the face of tooth.

For the flank or root, two lines, *a* and *b*, are drawn from the *side of the tooth*, also forming a right angle, *b* being radial; by the aid of the line *a* and the "setting" for the flank, the Odontograph may now be set for the flank.

When a setting is found for one part of a tooth, the instrument should be screwed on a radius rod, which is moved around a center pin in the hub, so that in this Odontograph we have a ready-made templet for the teeth, and are not troubled with getting centers for the points of compasses somewhere outside the wheel or between the teeth, as sometimes happens.

While it is very desirable to make gears that are interchangeable, and which are good enough for all ordinary machinery, it must be acknowledged that the best form of tooth cannot be made by any interchangeable system; of course, I am speaking now of the epicycloidal form of tooth. For some special machinery, where it is necessary to have the best form of teeth, without regard to their being interchangeable, the Odontograph is set differently from that for the interchangeable. Prof. Robinson gives extra tables for this purpose, and herein lies the value of the Templet Odontograph, and I would recommend its being used, especially for coarse pitches.

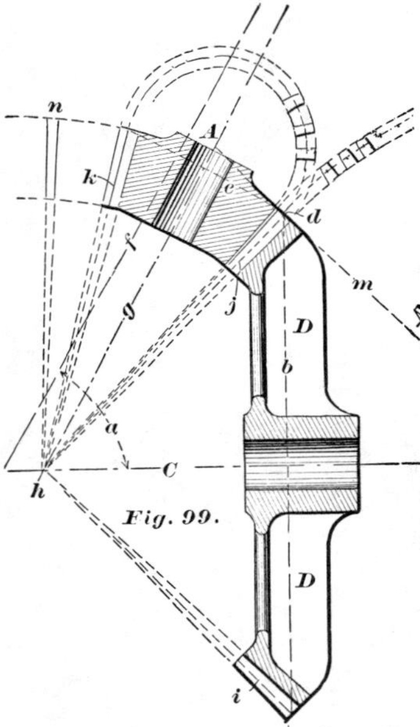

Fig. 99.

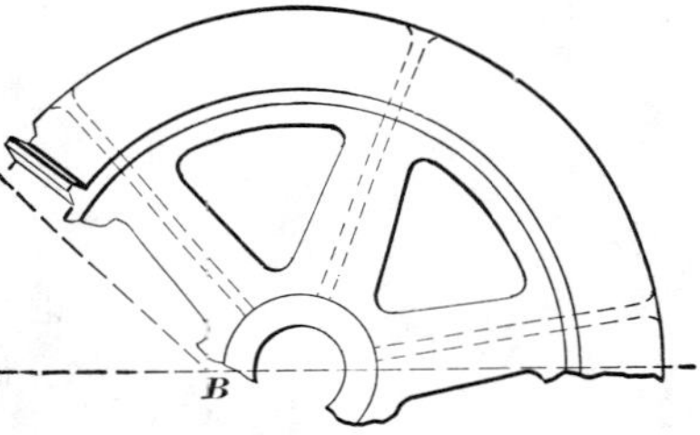

Fig. 100.

Fig. 104.

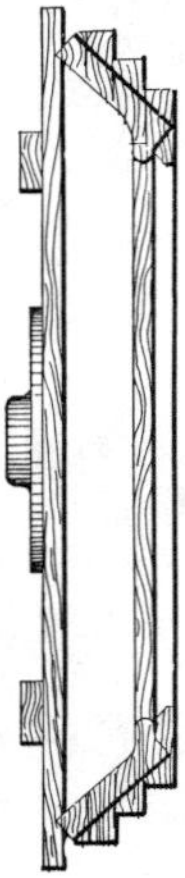

Fig. 101.

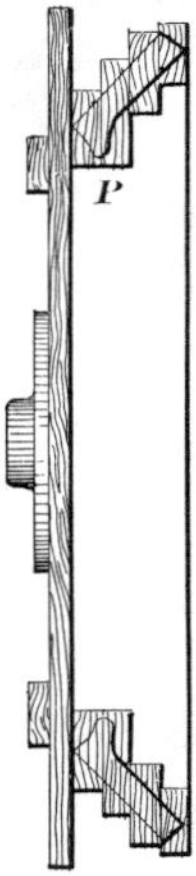

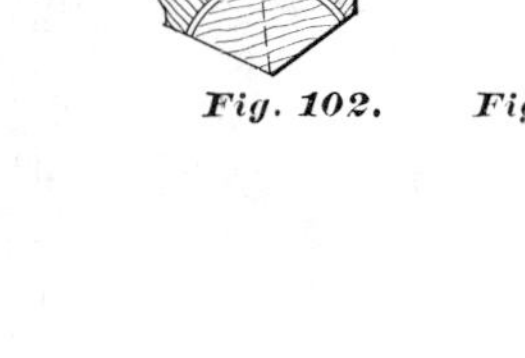

Fig. 102.

Fig. 103.

BEVEL GEARS.

BEVEL GEARS.

THE MANNER OF LAYING THEM OUT.

Before saying anything about the construction of these patterns, some explanation should be made about the lines that are required.

Fig. 99 is a section of the gear and pinion to be made, and gives the angle at which the two shafts are set. Fig. 100 is a face view of the gear. Bevel gears are usually made to run at right angles to each other, but when occasion requires, they can be made to run at any angle. In Fig. 99, *a* is the angle that is chosen; the two lines representing this angle are the first to be drawn, for on these all the others depend.

Having the angle, the next to be determined are the proportion and sizes of gear and pinion. In this case, I have made the proportion about 3 to 1, pitch 1½″. The gear is 26.8″ diameter, having 56 teeth, the pinion 9.11″ diameter, with 19 teeth. The pinion is made with an odd number of teeth, so that the teeth shall not work into the same ones of the gear at every revolution, but shall be constantly changing; the odd tooth is sometimes called a "hunting tooth."

The angle and the dimensions being settled, proceed by drawing the line *b*, at right angles to *c*. This line *b* is the diameter and the pitch line of the gear. From point *d* lay off *e* at right angles to *f*; this will be the pitch line and diameter of the pinion. Draw the center line, *g*, parallel to *f*; this locates the center, *h*, towards which all the teeth must converge. Draw *i*, *j*, *k*, the center lines of teeth, to *h*; on these lines lay off the face of the wheels, making the ends of teeth square with the center lines, *i*, *j*, *k*.

The section of the rim, arms, and hub, are now easily drawn. The teeth are laid out on larger circles than the diameters of wheels, the centers being *A* and *B*. These centers are obtained by producing the line representing the ends of teeth, or, in other words, making *m* at right angles to *j*, until it cuts the center lines of gear and pinion at *A* and *B*. The profile of the teeth must be made as if the centers of the gear and pinion were at *A* and *B*. To get the correct thickness of the tooth on the inside, the outer thickness must be laid down at *n*, running the sides toward the center, *h*. The profile of the inside of teeth is made after the same manner as the outside. Having described the method of drawing out these gears, we can pass to the construction of the patterns.

The rim is built up on a face-plate in a manner shown in Fig. 101, and when the work is thoroughly dry, the inside should be turned, getting in around at *P* as far as possible. The arms are fitted in the wheel as shown in Fig. 102, which should be done before taking the rim off the face-plate. There is a thickness of about ¾″ put on each side of the arms to hold them together, also answering for fillets; see Fig. 103. Tongues should be inserted at the joints where the arms join at the center, thus making, with the pieces on either side, a strong job. It is the intention to leave the hub and ribs, *DD*, in Fig.

99, loose, so that they can be lifted with the cope, thus preventing the mould from tearing down. After fitting in the arms, the rim is chucked, as seen in Fig. 104, and finished on the outside ready for putting on the teeth. I have already described the manner in which gear teeth are put on spur gears, and these should be put on in a similar way.

Pinions may be built up hollow or solid; in this case it is not large enough to build hollow, so pieces must be glued together with the grain of the wood running with the axis of pinion, and large enough to allow the teeth to be cut out of same. For larger pinions, the pieces can be glued around the periphery, the grain of wood running the same way as the teeth, then turned off, and the teeth cut as if cutting them from the solid, as before.

HOW TO LAY OUT THE THREAD OF A WORM FOR THE PATTERN.

LET us suppose the pattern for a worm is to be made 4″ diameter, with a single right-hand thread 1¾″ pitch.

First, turn the pattern—which should be in halves—to the diameter, 4″, and to the required length, say 6″, see Fig. 106. Do not destroy the centers, because after the threads are cut, the pattern can be returned to the lathe, and the threads sand-papered on a slow speed; a much better job can be done at sand-papering this way than otherwise. After turning the pattern, wrap a piece of paper around it and cut to the exact length of circumference, and also the length of pattern. After doing this, take the paper off and lay it flat, as shown in Fig. 105.

The full lines which are laid off parallel to each other, and 1¾″ apart, represent the center line of thread, or pitch. After making these lines, take the paper and wrap it around the pattern again, and it will be found that the end marked 1 will meet the end marked 2, and 4 will meet 3, and 5 will meet 6, and so on, thus making one continuous line when wrapped on the pattern; but before gluing the paper to the pattern, mark the thickness of the thread. Be careful when you have a right-hand worm to make, that you do not make it left-hand by running the lines the wrong way; I have seen this done more than once. If it is to be a right-hand thread, the lines should run down towards the left-hand, as seen in Fig. 105. If it is to be a left-hand, they should run down towards the right.

When a double thread is required, instead of starting to draw the lines from the first division to the corner, start from the second, as shown by dotted lines. This will give a double thread, and, in fact, any number of threads can be obtained this way. If the angle of the thread is increased twice, two threads are the result; if three times, three threads, and so on until the number of threads are so many, and the angle so great, that we cease to call it a worm and begin to name it a spiral gear.

It will probably occur to some minds that if glue be used to fasten this paper to the pattern, the moisture of the glue will stretch it, so as to make the ends that meet lap over; this can be avoided and the ends need not be allowed to lap over, if a little care is exercised. Instead of spreading the glue over all the surface of the paper, put a little on the two ends that join, and also here and there on the surface; then lay the paper flat and roll the pattern carefully on it, the pattern gathering up the paper around it as it is rolled. If a spiral gear is to be laid out this way—and it can be—the two ends of the paper should meet exactly.

WORM WHEELS.

THE WAY TO GET THE ANGLE OF TEETH, AND THE MANNER OF FASTENING THEM ON.

A WORM-WHEEL pattern is not an easy thing to make by any means; it is that kind of a job on which a good man is apt to go astray.

Figs. 107 and 108 represent a wheel to be made; the worm is shown in gear, and as I have already referred to it, I will confine my remarks to the wheel pattern, except that reference will have to be made to the angle of thread of worm.

The dimensions are the first thing to be determined. Let the wheel be 21¾″ diameter, 1¾″ pitch, 39 teeth, form of tooth, involute, laid out with the aid of Prof. Willis' Odontograph.

Pattern makers generally shape the thread of a worm pattern the same as the tooth of the gear in which it is to work; but the sides of the thread should be straight, at an angle of 75 degrees with the axis of the worm. This is correct for an involute, because that is what a rack tooth would be, and the worm is similar to the rack in this case. A section of the wheel pattern is shown in Fig. 109; the rim is first built up and turned straight on the outside, the pattern being parted on line *AB*.

A little thought will have to be exercised in building and turning the rim so as not to do any unnecessary chucking. In the first place, the face plate for turning it

Fig. 105.

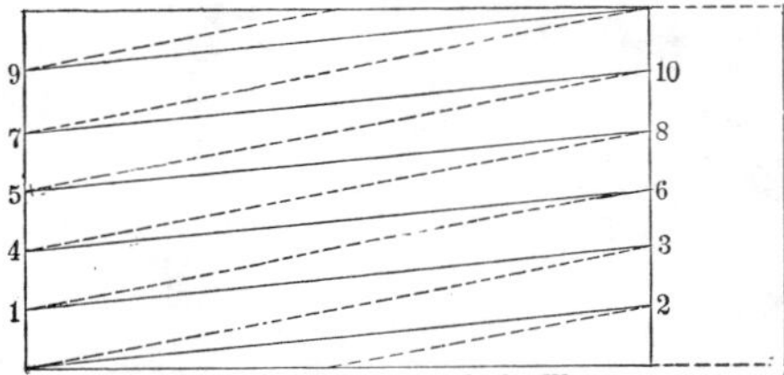

How to get the Lines for Thread of a Worm

Fig. 106.

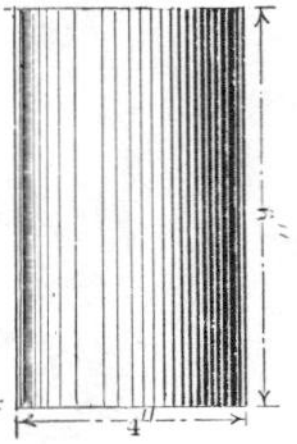

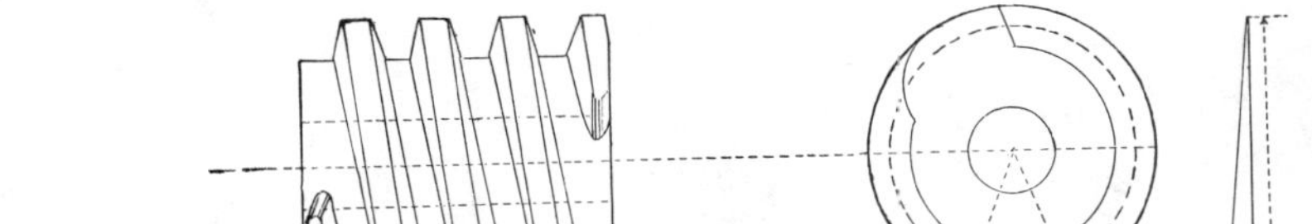

Fig. 110.

WORM WHEEL PATTERN.

should be about the same diameter as the rim, so as to allow the ends of the teeth to be turned off. Lay each half up separately, with the parting joint of pattern against the face plate, rough off the outside and finish the inside of both halves, care being taken to turn the inside of both halves the same diameter, because of the chucking. This done, the rings are chucked by the inside, and, to do this, segments are nailed on the face plate. The outside of the rim should now be finished to the size required, and blocks for the teeth fitted on the periphery. These blocks should be fitted and glued on with the grain of the wood running at the same angle that the teeth are to be at the pitch line, the width of each block being the same as the pitch. The way to get the angle is shown in Fig. 110, *y* representing the angle of tooth at the pitch line. These blocks are fastened on before taking the first half out of the lathe, and a groove turned on the joint for locating the halves concentric with one another. When the second half is chucked and turned and the blocks for teeth fixed on, as in the case of the first half, a projection is turned on the joint of it to fit the groove on the joint of the other. The two halves should now be put together, and the blocks for the teeth turned off on the ends, and finished, as seen in section Fig. 109. A good surface for marking out the teeth may be made by varnishing these blocks with yellow varnish.

In Fig. 107, a part of the pattern is shown cut through line *AB*, Fig. 109; the other part, with the three teeth represents the outside and ends of the teeth. To be theoretically correct the teeth should be thinner on the ends than on the parting line, *AB*, but in practice, this is not generally considered, because it would make it difficult to draw the pattern if made that way; the teeth in cast gears are therefore made the same thickness on the line *C* as on the pitch line *D* in Fig. 107.

The Odontograph for locating a center for marking out the teeth is shown at *E*; the instrument is nothing more than an angle of 75° divided off in ¼″ spaces on one side for 4″ in length, the ¼″ spaces being subdivided. Zero on the instrument is placed at a point on the pitch line, and the plain side is set to a radial line; the radius of the gear is then read off on the graduated side, which will locate a center from which to strike the tooth. In this case, the wheel being 21¾″ diameter, the radius is 10⅞″; read off 11, which is near enough for all practical purposes.

To lay out the teeth on the pattern at the proper angle, space off half the pitch from one of the joints of the blocks, on the pitch line each side of the wheel. If these pieces for the teeth have been made to the angle of *y*, Fig. 110, these points on each side will be the starting points for dividing the teeth, and will give the angle of tooth.

If the teeth are marked out also on the joint of one-half of the pattern, and trimmed down to the line, it will be an additional guide to cut the teeth by.

The fact of the arms of the pattern being in halves makes them thin, and therefore weak; but if the joints in the center be tongued, and the hub on each side glued and screwed to the arms, it will strengthen them considerably.

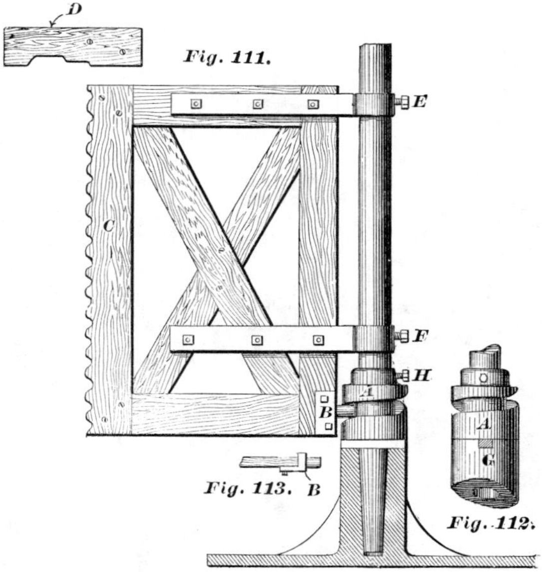

SWEEPING DRUMS.

SWEEPING STRAIGHT WINDING DRUMS.

FIG. 111 shows one way of sweeping up hoisting drums. The usual way is to allow one or two fingers for sweeping the groove to travel the whole length of drum by means of a nut running on a long screw. It will be seen that the long screw is discarded and that the sweep is made the whole length that the drum is to be. This sweep is required to travel up and down only one thread, which will make a continuous thread, the same as if two fingers traveled the whole length by means of the screw. The set screws, *E* and *F*, are free from the spindle, so that the sweep may rise and fall as it is pulled around. It rises by means of a pinplate, *B*, bolted to the bottom of the sweep, and which works in a spiral groove cut in the hub, *A*; this groove must be cut the same pitch that the groove of drum is intended to be.

Figs. 112 and 113 are other views of *A* and *B*. The spindle is kept from turning by the set screw, *H*, being set against it; at the same time, the projection on the hub, *A*, fits into the slot *G*, in Fig. 112.

After sweeping the grooved part of drum, the part *C* of the sweep may be taken off and a piece, like *D*, fastened to the upper cross-piece; this is for striking the flange of drum and a step around the top for a guide by which to set the cope. While sweeping the groove the spindle remains stationary, but when sweeping this flange and step, it must revolve with the sweep; to do this, the set screws, *E* and *F*, must be tightened, the pin-plate, *B*, taken off the sweep, and the set screw, *H*, loosened; this will allow the spindle to revolve in its taper socket in the ordinary way.

For drums over 6 ft. in diameter, it will be found necessary to support the top of the spindle with a temporary cross-beam, and also, to have a counterbalance weight to help the rising of the sweep as it is pulled around.

Fig. 115.

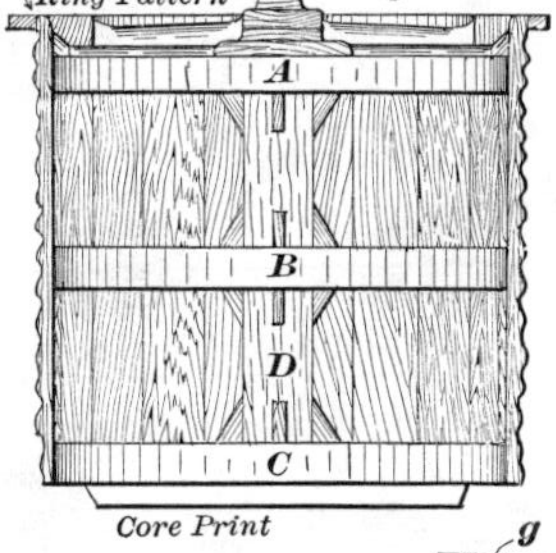

Fig. 116.

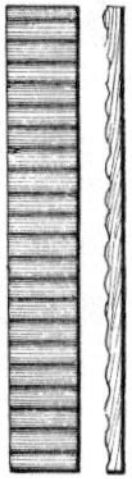

Fig. 114.

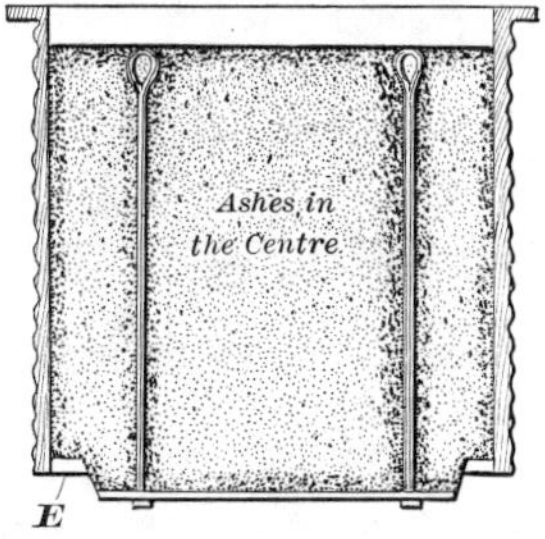

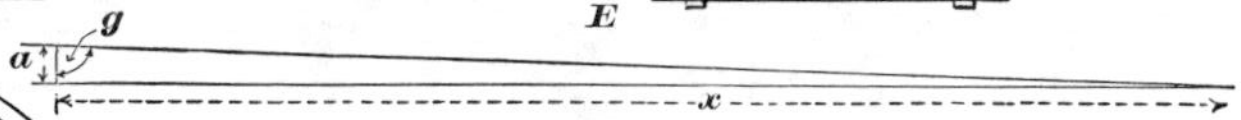

Fig. 117.

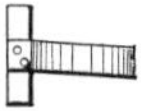

Fig. 118.

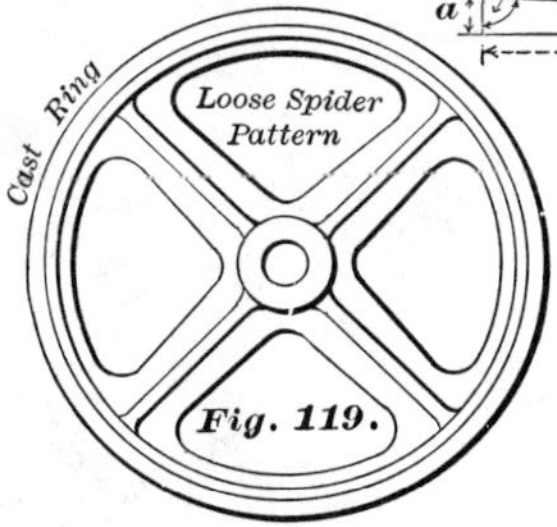

Fig. 119.

PATTERN FOR WINDING DRUM.

MAKING WINDING DRUMS FROM PATTERNS.

METHOD OF CUTTING THE GROOVE.

FIG. 114 shows a section of a Grooved Winding Drum about 3 ft. diameter, and Fig. 115 shows the way pattern may be constructed.

D is a 4x4″ stick; on it three discs, *A*, *B*, *C*, are fastened and screwed by brackets. Lagging to form the shell is screwed around these discs, while the spider in the end is loose and just laid on disc *A*, so that it can be lifted with the cope. *C* should be made about $\frac{3}{4}$″ smaller in diameter than *A*, to allow the core, shown in Fig. 114, to be easily lifted; this $\frac{3}{4}$″ taper on the drum will not be noticeable when cast, neither does it affect anything.

Fig. 116 represents a piece of the lagging that is screwed on the discs, *A*, *B*, *C*. Fig. 117 is given to show how to get the proper angle of the groove; the distance, x, is the circumference, a the pitch, and y the angle that the grooves are to be cut. Fasten two pieces together in the form of a **T** square, like Fig. 118, for marking the lines on Fig. 116, and let the blade be circular, to fit the face of the lagging. Also make a templet the same length and thickness of the drum, to mark off the grooves on the edge of lagging.

As the outside is being rammed up, the screws that hold the lagging must be taken out, so as to allow *A*, *B*,

C, and *D* to be removed after the cope is lifted. Previous to ramming up the core, a flange, *E*, Fig. 114, is placed in the bottom of the mould; this flange is made in segments and is for bolting the brake wheel to the drum. After the inside is rammed up and lifted out, this flange, *E*, and the lagging are drawn in and the groove finished off and the skin dried.

Some may say that a pattern such as I have described is very costly; in reply I would say that it would soon pay for itself by the difference in the cost of sweeping and moulding, but drums of larger diameter should be swept up, as they generally are.

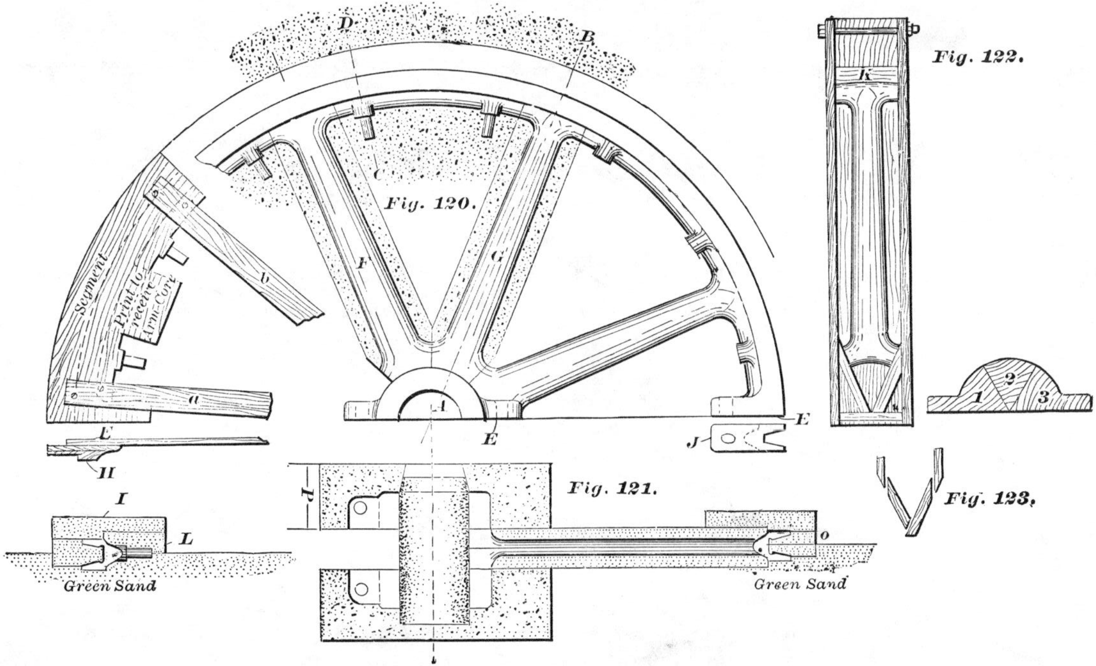

MAKING SHEAVES FROM CORE BOXES.

MAKING SHEAVES FROM CORE-BOXES.

I PROPOSE to give under this head some ideas for making sheaves of various kinds, and will first give the way for making large ones from core-boxes.

The style of sheaves, shown with cast arms, is that which is used to transmit power on cable car roads. The groove is made to receive segments that are bolted on, and of which more will be said further on. The right hand side of Fig. 120 shows one arm and a portion of the rim; as will be seen, this sheave is in halves, and arranged for bolting together; joint *E* is to be finished so that stock for planing must be allowed at *E*. In Fig. 121 two sections are shown; that on the right is a sectional view of the mould through the arm-core, *AB*, and that on the left through *CD*. These two sections represent the mould closed, but in Fig. 120 I have shown the mould open, with the arms, *F* and *G*, set. The lower part of rim is made in green sand, a segment pattern being used to form it. This segment pattern is shown to the left in Fig. 120, *H* being a section of it. Two battens, *a*, *b*, are screwed on, which run to the center, the ends being cut to fit around the spindle.

The lugs at the joint, for bolting the rim together, must be right and left hand on the segment pattern, using one when starting and the other when finishing. These lugs are screwed on temporarily. Care must be taken to divide the arms off accurately on a level bed. After marking the center line on the bed for each arm, move the segment around on the spindle and apply the center line of the core print for the arm core to the center lines on the bed, and, as lines on sand soon disappear, it will be well to drive a small stake on each side of the print for arm core, so that, when moving the segment to each division, the arm print can be set down between the stakes, and thus insure accuracy. It is particularly necessary to divide these arms off equally, so that the bolt-holes shall match those in the segments that go in the groove, for it is the intention that all the holes, both for bolting together the sheave and segments, shall be cast in, and to compensate for any difference that may occur, the holes are made a little long in the rim and lugs, as seen at *J*, Fig. 120.

Fig. 122 is the arm core-box, and I will again remind the pattern maker that it should be made strong, or it will come apart, as I have often seen, and then there is trouble about the cores not coming together at the center as they should.

Fig. 123 shows the way some pattern makers construct these large arm core-boxes; the result is, they are rammed apart, as shown, and then the poor core maker is accused of using the core-box roughly. Just as in the case of arm box for fly-wheels, so this box should be made a little longer than is necessary for the present job, so that the end, *K*, can be moved out and the arm lengthened for a larger wheel whenever it may be needed. The interchangeable pieces, 1, 2, 3, form the hub.

Fig. 124 is the core-box from which the cope cores, *L*,

Fig. 124.

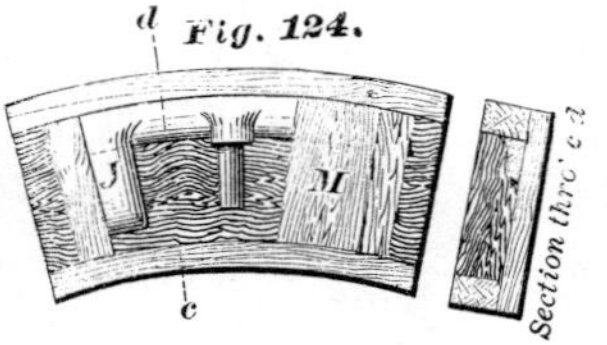

Fig. 125.

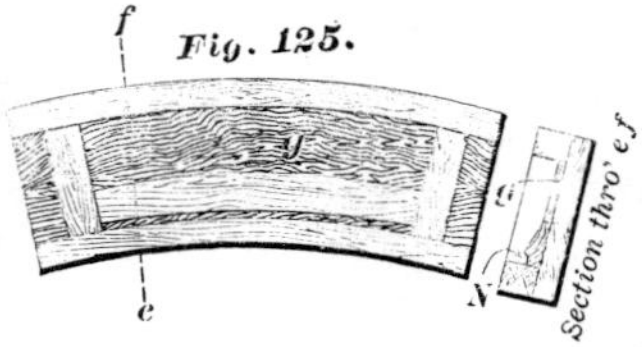

Fig. 126.

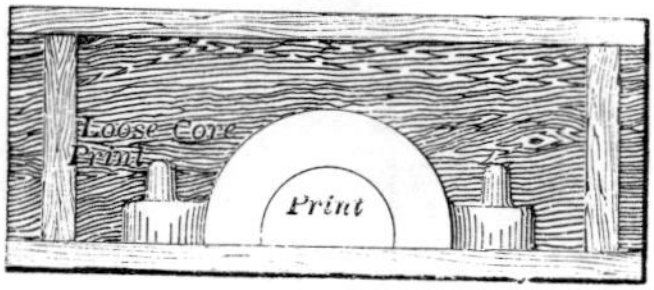

MAKING SHEAVES FROM CORE BOXES.

in Fig. 121, are made. The same lugs, *J*, that are used on the segment pattern, can be used in this box for two cores, using one right hand and one left hand. The box is shown arranged for the first core on the left. *M* is a loose piece half the width of the core print that receives the arm core. It will be seen that this box is made longer than the core is needed. This is to enable us to change ends with the loose piece, *M*, when making a core the opposite hand. The length of this box is from the center of the arm to the joint *E*, but the $\frac{1}{4}''$ stock, which is allowed for planing the joint at *E*, makes the box $\frac{1}{4}''$ longer than the eighth part of the half sheave, and therefore $\frac{1}{4}''$ too long for the other cores, so that a $\frac{1}{4}''$ piece must be put in the end of the box after making two cores with the lug, *J*.

Fig. 125 is the box for the groove cores. The section of this box shows it arranged for making the cores in two parts, to be pasted together at *o o*, Fig. 121. The cope part of this core is a little different from the bottom part; a loose piece, *g*, in Fig. 125, is fitted in the bottom of the box, to be left in for the bottom part of core and taken out for the cope part. The groove that this core is to form is turned on the two sides, but not in the bottom; tool clearance should therefore be allowed on each side in the bottom to accommodate the turning. This is seen in the section at *N*.

I have not shown any core-box for the slab core, *I*, as it is nothing but a plain core, and the box is simple and needs no explanation as to making it.

Fig. 126 is the core-box for the hub. It is made the depth of *P*, in Fig. 121. This view gives all the explanation that is necessary as to the way to make it. The faces, *E*, of the hub and lugs are covered with slab cores.

Another type of sheaves is shown in Fig. 127. This style can be made considerably cheaper than those I have already described. Dispensing with the groove that receives the segments makes it very simple to mould, and also easier for turning the periphery of casting. The plan for forming the arms and the hub is the same here as in Fig. 120. The segment pattern for making the green sand part of mould is seen in Fig. 128, of which Fig. 129 is an end view. The pieces that run to the center are not screwed on top of segment in the usual way, but on the step that is made in the segment at *a*, Fig. 129. By making it in this way, a large part of the rim can be made in green sand, as shown in Fig. 130.

The core-box for core *A*, is made in very much the same way as in Fig. 124. The segment that is bolted on the periphery of this sheave is moulded edgewise; there are chipping strips on the inside at *b*, *c*, *d*, *e*, Fig. 131; on the side, *f*, stock for planing is allowed, so that the segment may set straight against the flange of the wheel.

I once had a little experience with some of these segments. When the first lot of those in Fig. 131 were being fitted on the wheel, it was found they had straightened somewhat, just as represented by the dotted lines; it was evident that these segments straightened in cooling, the two thin flanges, *x x*, cooling first and pulling the casting out of its true circle; in the next, I took care to allow for this when making the pattern.

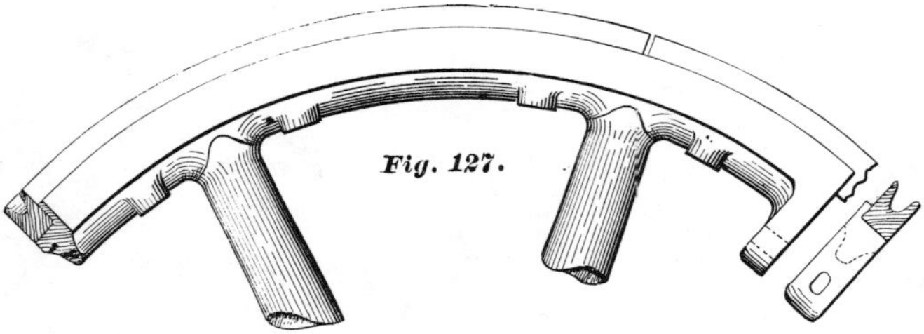

Fig. 127.

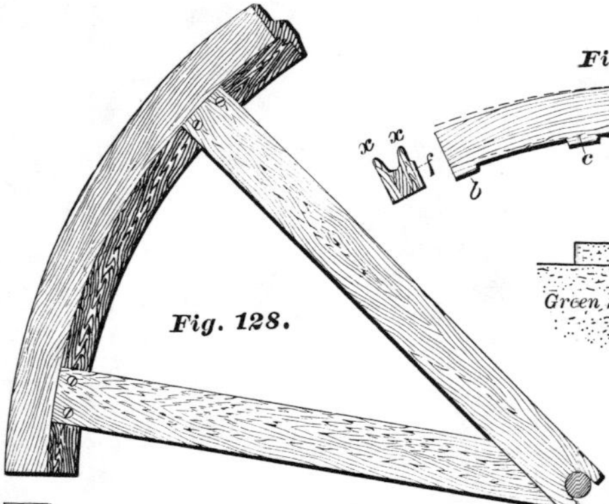

Fig. 128.

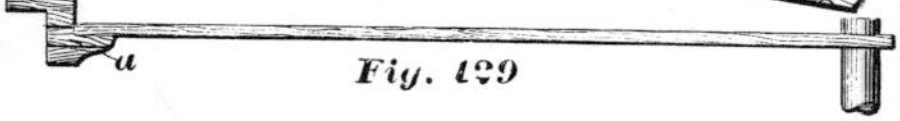

Fig. 129

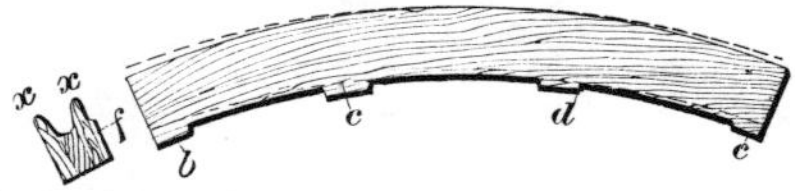

Fig. 131.

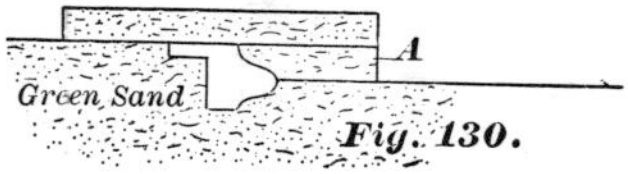

Fig. 130.

MAKING SHEAVES FROM CORE BOXES.

When there is a number of these sheaves to make, instead of closing the top with slab cores, as I have shown, it would pay to make a cast iron half ring with which to cover the top. This half ring should have a number of spikes on one side, and on it a thick coat of loam, struck off level, dried, and blacked.

MAKING SHEAVES FROM PATTERNS.

THERE is not much scheming required to make a pattern for a sheave, such as shown in Fig. 132, and yet, to show the way it should be made, may not be entirely out of place here, as I want to bring in a few points that have not hitherto been considered.

I have said that into the groove of this style of sheave are bolted segments that take the cable. The advantage of this arrangement is evident, as it allows the segments to be renewed when worn out. I have shown in Fig. 132 a part of the rim and a cross-section of the sheave; this shows the manner of bolting the segments to the sheave. The groove, into which the segments are bolted, is to be turned, but the groove of the segment is left rough.

Chipping pieces are cast on each side of the segment, as seen at *a, b, c, d,* Fig. 133, because it is intended that the segment shall not bear in the bottom of the groove, but only on the chipping pieces by the sides, and at *e e*; see cross-section, Fig. 132.

Fig. 133 shows the pattern of the segment, and is made to be moulded on the edge, the groove being in the cope; it is desired to cast the bolt-holes, and care must be taken in spacing them off, because they are wanted to match those in the sheave, which are also cast in. I have marked the core prints for these holes; the bottom print is made something like the cope print—

Fig. 132.

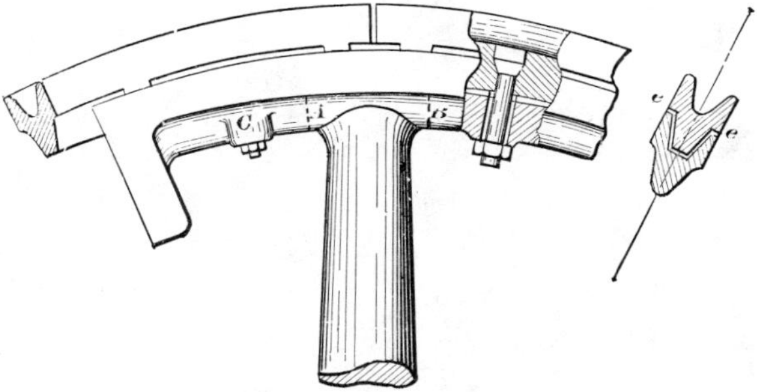

Fig. 133.

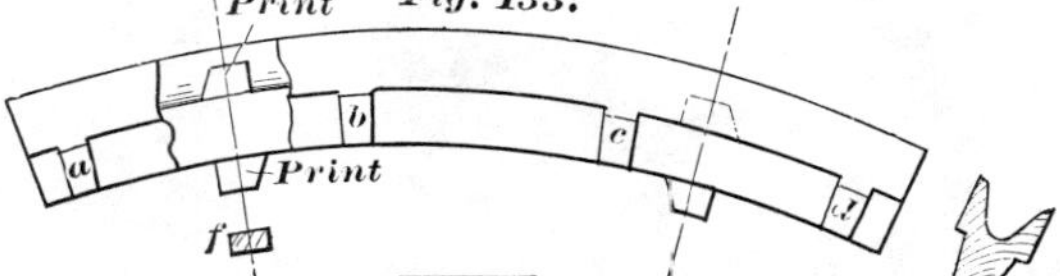

Fig. 134.

Fig. 135.

Fig. 137.

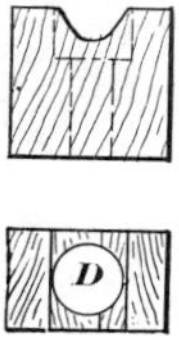

Fig. 136.

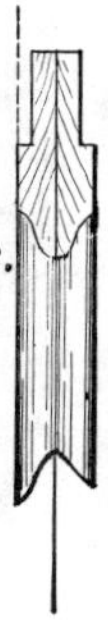

MAKING SHEAVES FROM PATTERNS.

oblong—as shown at *f*; this is done in order that the core may stand in the mould more securely while the cope is being closed. If a round print were used just the size of holes, the cores would be top-heavy and difficult to locate in the mould, hence the necessity of making the bottom print as shown. The core-box for this bolt-hole should be made as shown in Fig. 134.

It is understood that these sheaves are bolted together in halves, so that in making a pattern, only one-half will be required. Proceed by building up and turning a whole ring, of which Fig. 135 shall be the section; *A* is the print for carrying the groove cores. After turning the ring, cut a stick the exact length of the inside diameter—this will be a gauge to see whether the ring has sprung after being cut in two, and, if it has, to bring it back to the gauge when fastening in the arms. After sawing the ring in two, glue and screw it together strongly, as seen in Fig. 136; but before doing so, it must be remembered, as before, that stock for planing must be allowed at the joints where the ring is bolted together, so that the pattern shall be $\frac{1}{4}''$ over the half circle. In order that this may be, the ring should not be cut exactly in halves, but $\frac{1}{4}''$ one side of the center, making one part about $\frac{1}{2}''$ short, not reckoning anything for saw cut—with saw cut would probably be $\frac{3}{4}''$ short. Now, if after sawing the ring, two of the ends be brought together, it will only be necessary to build on one end of one of the sections. Having done this, the ring is ready to have the arms fitted in, which should be done by letting them in the rim, as represented by dotted lines at *A B*, Fig. 132.

Care must be taken not to fit the arms in so tightly as to spring the ring out of round; this can be very easily done. After locating the arms, bore two $\frac{3}{4}''$ holes from the outside of ring into each arm, and glue in hard wood dowel pins; this will make a strong job. The small bosses, of which one is shown at *C*, Fig. 132, are turned and sawed out with a narrow band-saw to fit over the rim. This is done by inserting each boss in a block with a hole through it the size of the boss; two views of this block are seen in Fig. 137. The boss is fixed in the hole *D*, and sawed to the shape of the inner part of rim. Of course, the block is fitted over the rim first, to act as a guide for sawing them out.

The core-box for the groove need not be made with loose piece in the bottom, as in case of forming these sheaves with cores, because the cope closes down on top of print, and not on the dotted line, Fig. 136.

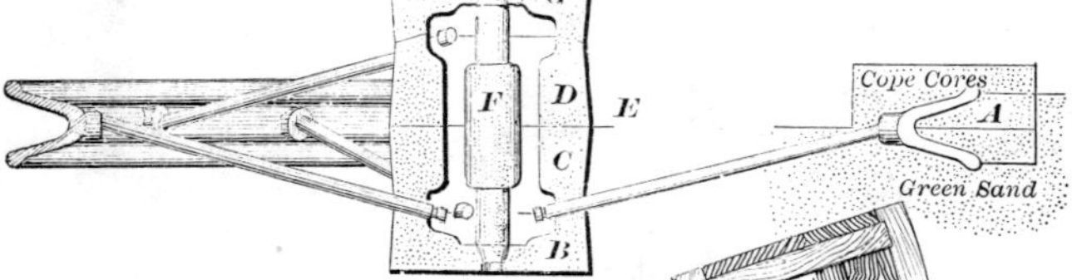

Fig. 138.

Fig. 144.

Fig. 145.

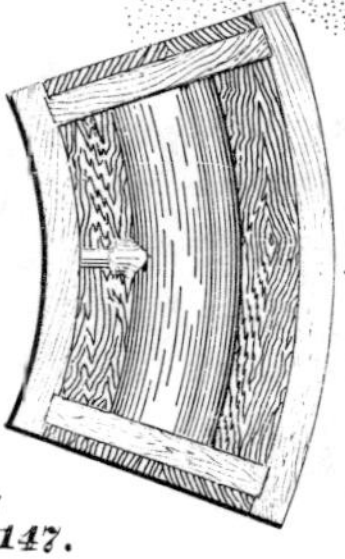

Fig. 140

Fig. 146.

Fig. 147.

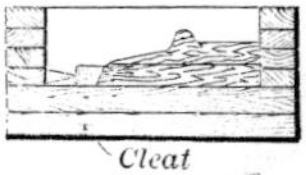

Fig. 141.

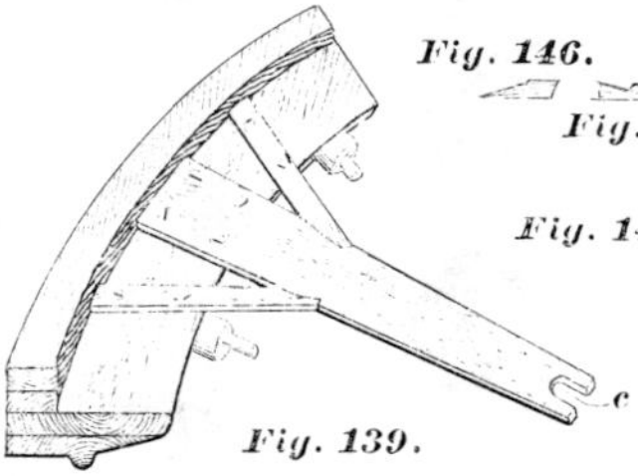

Fig. 139.

Fig. 142.

Fig. 143.

SHEAVES WITH WROUGHT IRON ARMS.

SHEAVES WITH WROUGHT IRON ARMS.

AN ORIGINAL WAY OF MAKING THE HUB.

THE style of sheave shown in Fig. 138 is used extensively in mines for carrying rope; the arms, which spread on either side, act like stay-rods to the rim, making it very rigid sidewise, at the same time forming altogether, a light, but strong, sheave.

To the left of Fig. 138 is shown a section of the rim with wrought iron arms cast into it; to the right, a section of the cores which form the rim; and at the center, a section of the cores forming the hub.

The lower part of the mould is formed with green sand, the segment shown in Fig. 139 being swung around from the center, *C*. The cope is formed with cores made from box shown in Fig. 140. Fig. 141 is a cross-section of this box.

While this is a good way to make sheaves of large diameters, for those under 8 ft. diameter a full ring is probably a better way, providing the ring can be stored so as to lie flat on its side, instead of standing on its edge; for, having no arms, a large ring standing edgewise would soon become oval.

The cores forming the groove are made in halves from the box, of which Figs. 142 and 143 are two views; these cores are pasted together at the joint, *A*, Fig. 138. Four round cores are made to form the hub; these are set one on top of the other after making the lower part of the mound with the segment. The lower hub core, *B*, through which there is a hole, should be set over the pin from which the segment has been swung around. This will locate the hub concentric with the rim. Half of the arms should now be set in the mould, after which, the two middle cores, *C* and *D*, are located. When *C* and *D* are being pasted together at the joint, *E*, care should be taken to get the holes that receive the arms exactly midway between those in the lower part. The center core, *F*, should now be set, then the balance of the arms and the top core, *G*. When pouring these sheaves, the rim is allowed to shrink all it will before pouring the hub, and in large ones, the hub is not poured until the following day.

Figs. 144 and 145 are sections of the round core-boxes for the hub cores, *CD* and *DG*. Plenty of draught should be made on the inside of these two boxes at *a a*; the half round prints for the arms are shown at *b b*.

The small bosses shown in Figs. 146 and 147 are used in cope core-box, Fig. 140; the prints on these small arm bosses vary; the cores which cover those arms running upward from the hub, should have Fig. 146 in the box, and those which run downward, Fig. 147. The bosses on the segment are made similar.

Fig. 148.

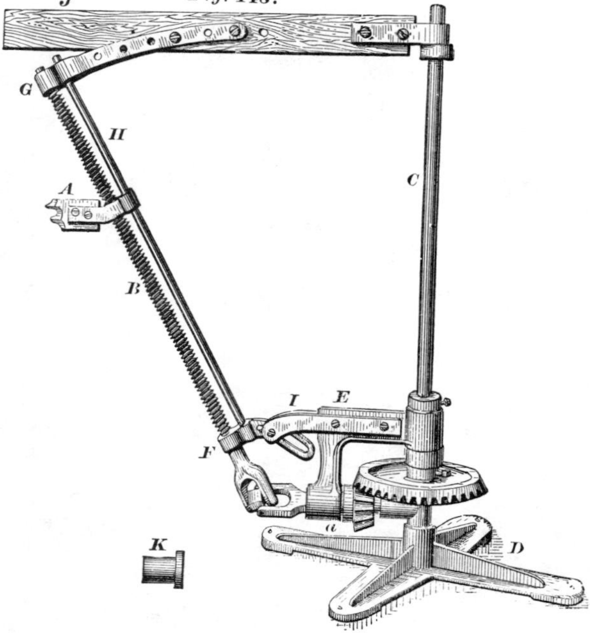

SWEEPING CONICAL DRUMS.

A MACHINE

FOR

SWEEPING CONICAL DRUMS.

DESIGNED BY THE AUTHOR.

It may not be understood by some why a winding drum is sometimes made conical instead of a straight cylindrical form, and it may not be entirely out of place here to explain the reason, for the benefit of such.

Conical drums are used for winding heavy loads from deep mines. When the skip or load is at the bottom of the mine, ready to be hauled up, the winding on the drum begins at the small end, and, as the rope does not wind as fast on the small end as it does on the large end of the drum, it allows the slack rope in the shaft to be gradually taken up at the starting, and also prevents the load from starting too suddenly. The engines also gain a decided advantage when winding with conical drums, because, instead of the winding being started at full speed, it gradually increases, thus giving the engines a better chance to do their work.

It is scarcely necessary to inform my readers that it requires a great deal more skill to build and properly secure a mould for a large conical drum than it does to mould a grate bar; but there is a class that stands so high in the engineering profession, that to them all foundry work is just a little above unskilled labor—something requiring more brute force than anything else. Such ideas, though, do not prevail among our genuine and practical engineers; they are only found among the cleverly ignorant.

All those who have much to do with the machinery business know what an amount of consultation and scheming is necessary before some jobs in a foundry can be started, and then how it requires men of good sound judgment to execute the work.

The building of a mould for a large conical drum is one of these jobs. The way of sweeping the groove, an arrangement for which I propose to describe, is only a small item of the work.

In Fig. 148, *A* is the sweep that travels up and down the screw, *B*, as it is pulled around. The spindle, *C*, is secured to the cross, *D*, at the bottom; the bevel gear is fast on the spindle, two set-screws in the hub holding it in place; the bracket, *E*, is loose, and turns on the spindle; it has a bearing at *a*, in which the pinion shaft runs; the end of this shaft is carried by a tee piece that turns on the spindle. The pinion shaft and the screw are connected by a universal joint, while the screw is carried by two adjustable curved pieces, *F* and *G*. Guide-rod, *H*, keeps the nut from turning on screw, *B*; arm, *I*, fits over the bracket, *E*, and carries the curved piece, *F*; this arm is also adjustable.

Now, it will be clearly seen that, if the arm, *J*, and the bracket, *E*, are pulled around, it will cause the pinion and the screw, *B*, to turn, thus making the sweep, *A*, to travel a certain distance every time it goes around. The gears determine the pitch of the groove to be swept; if the proportion of the gears are three to one, and the screw $\frac{1}{2}''$ pitch, then the sweep will travel $1\frac{1}{2}''$ at every turn, making a groove $1\frac{1}{2}''$ pitch. When any other pitch is required, the gears must be changed for those of a different proportion, for instance, for a $2''$ pitch drum the proportion of the gears would be 4 to 1. Bracket, *E*, is made so as to permit the use of gears of different sizes.

When a drum is wanted with a left-hand groove, the gear on the spindle is turned upside down, and located under the pinion instead of over it. The machine is also arranged so that a drum of any angle can be swept. This is done by loosening the bolt that holds part *F* in place, and by taking out those in the upper part, *G*, thus allowing the screw to be swung at any angle from the center of the universal joint.

The reason for making the arm, *I*, separate from the bracket, *E*, is obvious; it is to give a better chance for adjusting the lower part of the machine than the swinging of the screw gives. The pinion shaft runs in close to the upright spindle, so that when the set-screw in the pinion is loosened, the shaft can be pulled out to the required distance, and the set-screw in pinion be tightened again. When this is done, it will be found necessary to bolt on the flanged sleeve, *K*, to the end of bracket, *E*, between the universal joint and the bearing, *a*. For building and sweeping up the mould roughly, the screw and pinion shaft can be disconnected entirely and a plain sweep made, bolting it to the upper and lower arms, *J* and *I*.

The engraving only represents the model which I

made of this machine; the details of the machine proper will vary somewhat. For instance, where there are solid boxes on the bracket, *E*, for the spindle and pinion shaft, there should be caps, so as to make it easier to disconnect the parts. The universal joint should also be made separate from the screw and pinion shaft; many other items would need changing when building a machine to do the work.

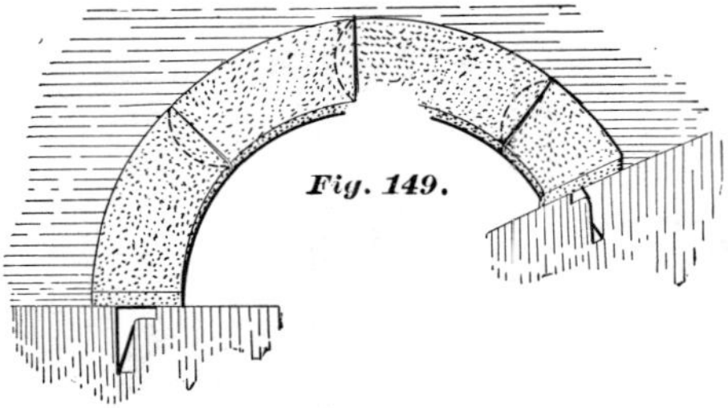

Fig. 149.

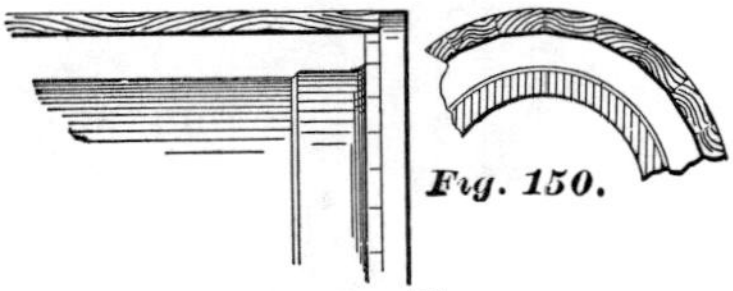

Fig. 150.

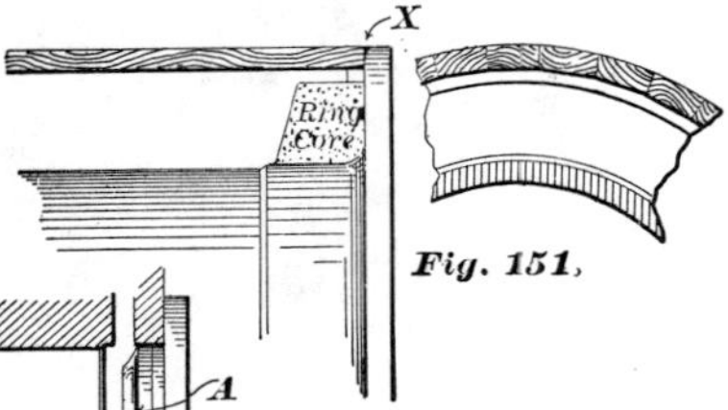

Fig. 151.

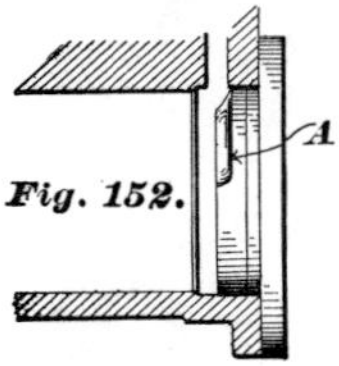

Fig. 152.

ITEMS FOR PATTERN MAKERS.

ITEMS FOR PATTERN MAKERS.

It is surprising the amount of time and money which may be saved in the machine shop and foundry when attention is given in the pattern shop to what is sometimes termed the smaller matter of the business. Usually the more important details do not wear us half as much as the ever-recurring little set-backs. For instance, how often it happens that the machinist has to chip away a cylinder head at *A*, like shown in Fig. 152, to prevent it from choking the port of the cylinder, when the pattern maker, might just as well have cut the clearance on the head back far enough to clear the port 1/8", and thus save the machinist's time. I do not think any M. E. would be fine haired enough to object to the small amount of extra clearance space that this would give.

Again, what unnecessary trouble might be saved sometimes in chucking a casting in the lathe, if some lugs or other projections, that could easily be broken off the casting, had been put on the pattern to facilitate the holding of it. I have often seen a machinist with a casting on the planer or lathe puzzled how to hold the "blamed thing," as he would name it, from jumping out of the machine, when perhaps a little foresight in the pattern shop would have made matters easy. A good pattern maker will consider these things and look ahead, following his job in imagination through the foundry and into the machine shop.

Another thing along the line of items, and which is generally overlooked: when a cylinder casting has to be lagged with black walnut, the pattern maker should make provision on the backs of the flanges for surporting the lagging, similiar to that shown in Fig. 150. This is a much better method than blocking the lagging off from the body of cylinder with blocks fastened on back of lagging.

The edges of the flanges at *x*, Fig. 151, should not be rounded off, as is often done. When flanges are extra wide, they might be cored out by setting a ring core on the backs of flanges, similar to that shown in Fig. 151.

By providing for the cylinder lagging this way, a better and cheaper job can be done than otherwise, when no provision is made.

Stamping or marking with black varnish the diameter and length of all core prints, on every pattern is very simple, and apparently unimportant, and yet if this practice were general, I fancy the moulders would pronounce an everlasting benediction on the pattern makers, especially the moulders who never have any calipers or rule—in fact, it would be an all-around convenience and a preventive of mistakes.

Many circular pieces of work in the foundry are covered with slab cores to save ramming up a cope, and a core-box for these is usually a very simple piece of work, and yet nine times out of every ten the cores when made and located do not fit on the ends, but will remain open, either on the inside or outside in the manner shown in Fig. 149. If the moulder should butt the ends closely together, the cores would not cover the mould, but would run off somewhere on the outside or inside, as the case may be. Now, I do not mean to say a pattern maker cannot make a core-box of this kind, and not make the ends radial, so that the cores shall fit closely together when laid down, but there is a very simple way by which the ends of these slab cores can be made to fit without any trouble to the moulder, and that is, by making one end round and the other hollow, as shown by dotted lines in Fig. 149. By this method the ends can be kept close together, and the cores located in their proper place when covering the moulds.

The amount of stock that should be allowed on a a pattern for finishing in the machine shop is a source of trouble to some pattern makers. Among the first questions that many will ask their foreman when given a new job, are, "How much stock shall I allow for boring?" "How much for planing?" and so on. It is almost impossible to establish any rule that shall be uniform as to the amount of extra stock that should be allowed—for while 1/8" is not enough for some jobs, for others it may be too much. Circumstances and a workman's good judgment can alone determine the proper quantity to allow.

A dry sand casting will not shrink or strain as much as a green sand one, and in large work the shrinkage is uncertain, so that before a pattern maker can intelligently allow the desired amount for finishing, he should know the conditions under which the casting is to be made, and it will help him in his conclusions to get acquainted with the style or manner in which the moulders are accustomed to turn out work, for in some foundries the moulders will put on all the stock necessary for finish, and a great deal more, without the pattern maker allowing any.

It makes all the difference in the cost of a piece of large machine work, which has to be turned or planed to correct measurement, whether the machinist has to take three or four roughing cuts or just one; whether he has to take off an inch and a half, where a quarter of an inch would have answered. Then, again, the pattern maker must not be too "skinny" about allowing stock, or he will give the machinist more trouble than by allowing too much—of the two evils, too much to take off a casting is rather to be desired than not enough. Insufficient stock on a complicated casting will make a machinist bless the pattern maker in a manner not too complimentary.

One-quarter of an inch (½-inch in the diameter) ought to be enough for boring out say a 20″ x 48″ cylinder, but if the same amount only was allowed on a 40″ x 72″ cylinder, the bore in all probability would not clean out to proper size; ⅜″ or ½″ would not be too much on such a large cylinder.

On some castings the stock allowed on the cope side should be more than on the drag side; for instance, a large crank disc, a Corliss cylinder and many other jobs.

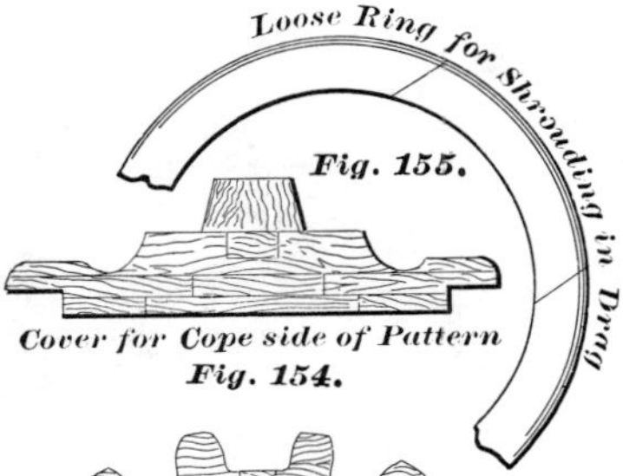

Loose Ring for Shrouding in Drag

Fig. 155.

Cover for Cope side of Pattern

Fig. 154.

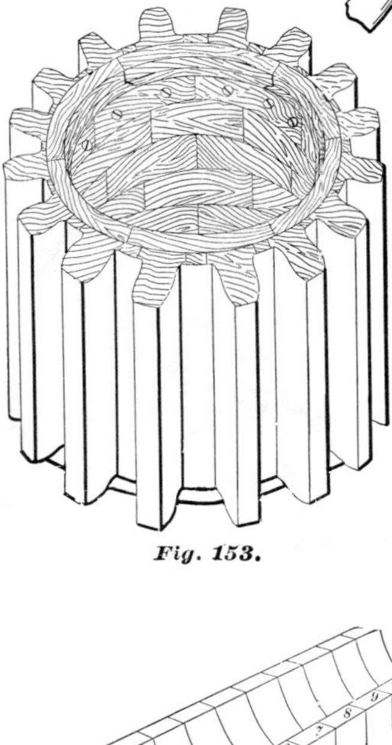

Fig. 153.

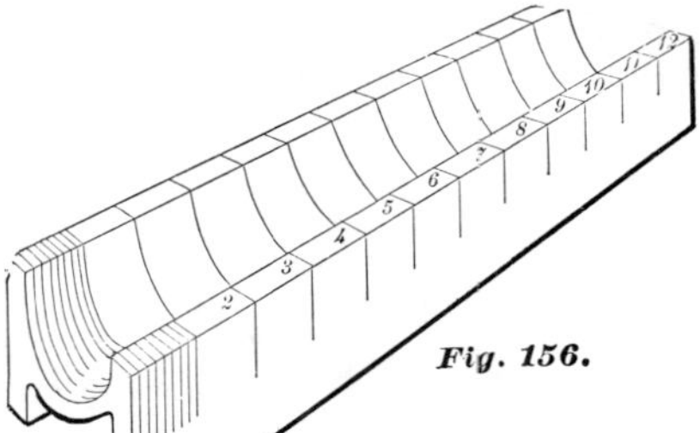

Fig. 156.

BUILDING UP SPUR GEAR PINIONS.

If spur gear pinion patterns, with about 7″ face and upwards, were built up hollow, with a cover on the cope side, which could be removed as shown in Figs. 153 and 154, the moulder would be more likely to get a casting like the pattern.

By leaving the cope side of the pattern open this way, it allows the moulder to rap all arouud on the inside of the pattern, which is a much better way of loosening the pattern than driving a spike in on the top, and hammering the spike sidewise, until the mould is enough countersunk to produce a bevel pinion instead of a spur pinion.

Building the pattern this way not only helps the moulder, but also makes it convenient for fastening the teeth on with screws from the inside, using a short screw-driver to do so.

For drawing the pattern, an iron plate with a tapped hole should be inserted on the *outside*, in the end that is moulded down, and a hole bored through the bottom—in this way the draw plate will not be so liable to be pulled out as if put on the inside.

The ring forming the shrouding over the teeth should be cut into segments in the manner shown in Fig. 155—in this manner the shrouding can be drawn in, thus doing away with the parting in the bottom of mould, which would have to be made if the ring were whole. The shrouding need only be carried to the pitch line of teeth.

FOUNDRY CORE-BOXES.

PATTERN makers are not supposed to make a core-box for every round core that may be required, except in cases of special sizes, or when chambered holes are called for. Foundries usually carry a stock of what are known as round cores of ordinary sizes, though it would be a libel on most foundries to call them round or say they are ordinary sizes.

I suppose the reason why foundry stock-cores have such a bad reputation is because there are so many cores constantly being made from one box, and the "artist in cores" becomes so attached to his "standard core-boxes" as to be unable or unwilling to discern when they are unfit for further service—no matter what condition the box be in, it will always stand to make a few more cores.

This is why castings come into the machine shop with a 2″ round hole (or rather what is supposed to be such), measuring all the way from 2″ to 2¼″—all depends what part it is measured.

A core-box that might do better service than the ordinary one is shown in Fig. 156. It can be made of iron or brass, according to size required. It will be seen that it represents a box 12″ long, spaced off in inches, a cut being made at each division with a hack saw, enabling the core maker to cut the core, while green, to the required length, by drawing a thin trowel or an old hack saw through one of the cuts. For an inch on one end it is spaced off in eighths—this gives the core maker an opportunity to practice "fine measuring" when cutting the core, but these eighths may be dispensed with, according to taste.

For large sizes the box should be made longer—how much longer depends upon the class of foundry work done. I am of opinion that if some enterprising firm were to make a line of half round iron core-boxes, something after the style shown in Fig. 156, and get them up light, they would find a ready sale for them in foundries all over the country; it would be meeting "a long-felt want."

MAKING AMMONIA PUMP CYLINDERS.

THERE are few firms manufacturing ice making and refrigerating machinery, which have not had some trouble at some time with their pump cylinders and other details, which have to withstand a high pressure.

When a casting is being tested, the leak, if any, appears in some corner of the casting, such as at the back of a flange where it joins the body—this is the most probable place for the defect to show itself, owing to the counter-shrinkage that takes place just at that point. A large fillet in the corner will sometimes help to prevent defects, but it is not always sufficient, especially where large flanges are cast on the body of a cylinder.

Pump cylinders for ice machines usually have large flanges, and to prevent the iron from drawing away in the corner at the backs of these flanges is one of the problems that the moulder has to solve.

I was in a foundry not long ago, where they resorted to what I consider a good way to prevent the iron from being too open in the corners, as, indeed, all castings are, wherever there is a point at which counter-shrinkage takes place. It was done by placing chills in the corner, at the back of the flange, the chills being rammed up with the pattern. This, of course, chilled the iron in the corner, making it closer grained than it otherwise would be, and it had the desired effect of producing good castings, that would stand the test without leaking.

PLUG VALVES.

Fig. 157.

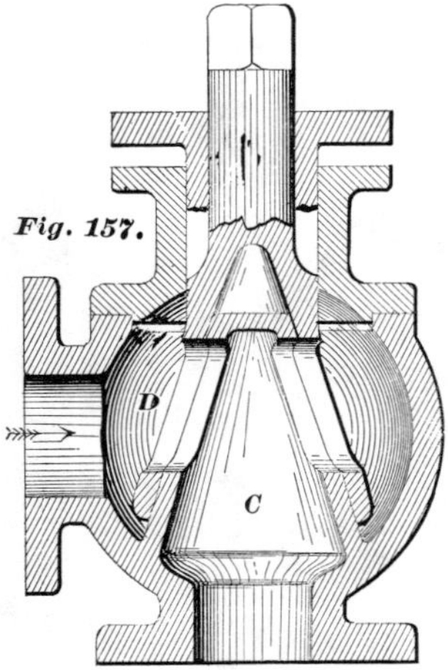

Fig. 158.

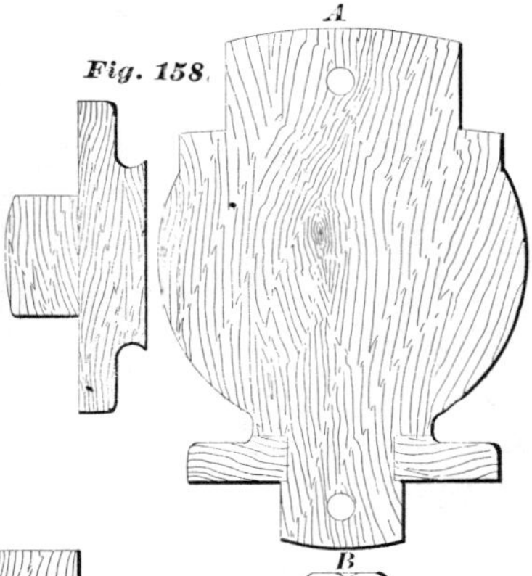

Fig. 159.

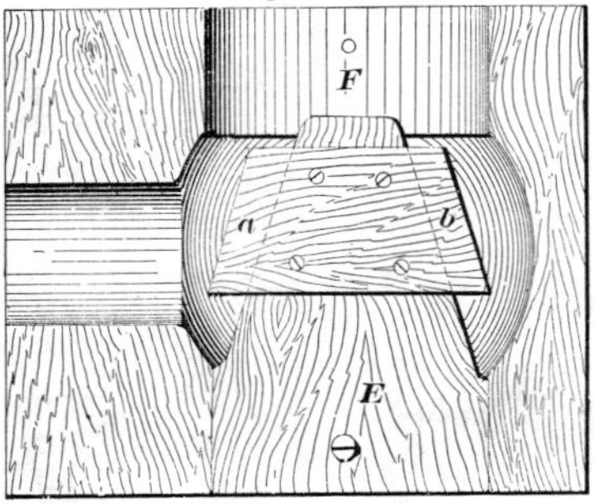

Fig. 161.

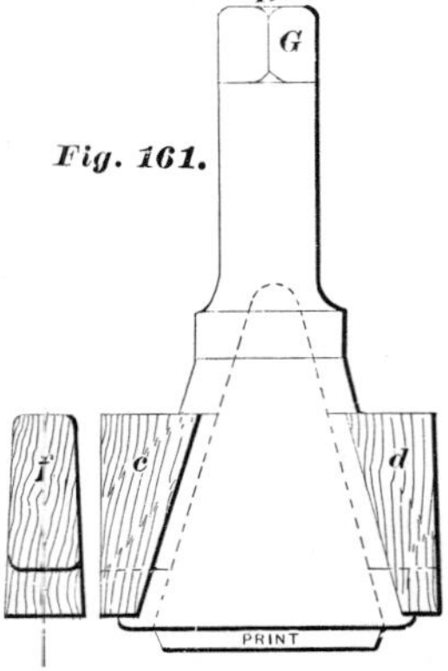

Fig. 160.

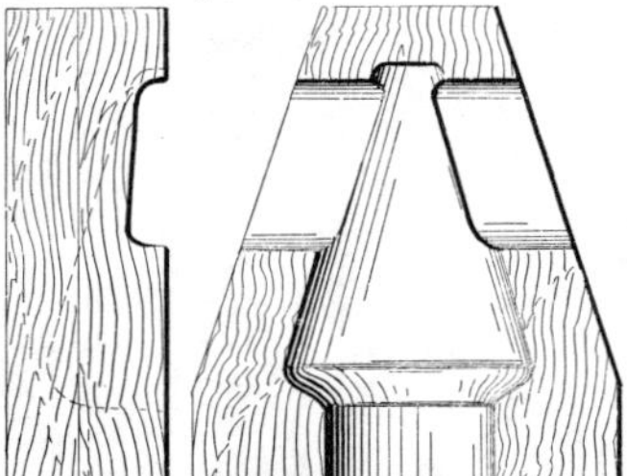

Fig. 162.

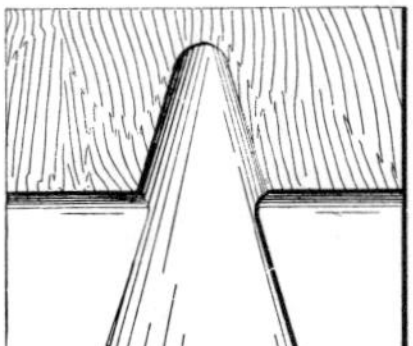

American Machinist

PLUG VALVES.

THERE have been so many improvements made in valves in recent years, that the ordinary plug valve is becoming almost a thing of the past. One of the reasons why many engineers have discarded the old forms is the liability of the plug or tumbler to fasten itself so tightly in the shell as to make it hard to turn. The cause for this "sticking fast," as it is generally termed, is not always the one so often given, viz., corrosion, but is rather the unequal expansion of the plug and shell, sometimes due to the bad proportioning of metal in the two parts, and by making the shell of cast iron and the plug of journal metal. The making of these two parts from different metals may prevent corrosion, but it rather increases the tendency to unequal expansion.

Fig. 157 is a section of a form of plug valve known in England as the "Etchell Tap," and is used mostly as a blow-off valve. The advantage claimed for this design over the old style is that it will not stick fast in its seat. The plug being inverted, a better distribution of metal is obtained, permitting the steam or hot water to circulate all around the inner parts of the valve, distributing the heat, by which means unequal expansion is obviated to a very large extent.

It is not intended under the above heading to discuss the merits of valves, so much as to show the manner of constructing the patterns and core-boxes; yet these few general remarks will not be out of place.

The pattern for the body of the valve is parted through the same plane as that of the section. Fig. 158 represents half of the pattern of the shell or body without the bonnet and stuffing-box. The ball part of this shell, and the two core-prints, are turned between its two centres *A B*, while the inlet piece is turned separately and fastened to the ball part, the latter being cut off flat to dotted lines for that purpose.

For large valves it is advisable to build up this pattern hollow, of segments, instead of making it solid, as shown, because a large solid block will shrink and crack, in which case it generally becomes necessary to plane off the joint of the pattern, to make the halves fit closely together again, as they should.

By referring to the section in Fig. 157, it will be seen that two core-boxes are needed for the shell, one for coring the space inside the conical part, and the other for the outside. Let these two spaces be disignated *C* and *D*. The core-box for the space *D* is shown in Fig. 159, and as two half cores made from this box would not match each other if pasted together, the box is arranged so that the cone *E* can be taken out and located at *F*, by which means half cores can be made that will match each other when put together; *a b* is a core-print let in flush with *E*—before rolling the core over on a plate, the cone *E* is drawn out and the space filled with green sand.

Fig. 160 is a core-box for the space *C* in Fig. 157, and is pasted in the main core *D* at *a b* when dry. The view to the left represents the side view of this box.

Fig. 161 is a pattern of the tumbler, or valve, as seen in the section in Fig. 157. This pattern is intended to be moulded on end with square part *G* down. The core, which is shown by dotted lines, is carried by the two core-prints marked *c d*. Notice that these two prints, of which *f* is an end view, are carried nearly to the end of the pattern. This, of course, is necessary for the sake of moulding, and for dropping in the core; but provision is made in the box at *x x*, Fig. 162, by which this part of the print is filled in by the core, so that the moulder is saved the trouble of building over the core.

It is unnecessary to show here the way of making the patterns for the stuffing-box and gland seen in Fig. 157, as they are so easily made that it would be uninteresting to ordinary pattern makers.

GOVERNOR VALVES.

THE two sections in Figs. 163 and 164 represent a valve that is not only used on governors, but is very often used in water tanks to regulate the supply of water, a hollow ball that floats being attached to the valve lever in the usual way.

In Fig. 163 the rotating valve *a* is shown in place with the two passages almost wide open, but in Fig. 164, which is a section of Fig. 163, through *A B*, the valve is removed, and is represented in Fig. 165.

There are ways of making the pattern and core-boxes for this style of valve other than are given here, but the method represented by the illustrations has been chosen because it is the way the writer would adopt.

Fig. 166 is the pattern which is parted on line *C D*, or, in other words, the joint of the pattern is the same as the outline of the section, Fig. 163. There is nothing to prevent the pattern from being parted on line *A B*, that is as Fig. 164 shows, but this would necessitate setting the cores in the mould in not quite so good a way as they can be, by parting the pattern on *C D*, and moulding it as represented in Fig. 166. Not much of this pattern can be turned in the lathe, but the flanges, prints, and a small part of the inlet and outlet, therefore the part which surrounds the valve chamber, will have to be worked off by hand.

Fig. 163.

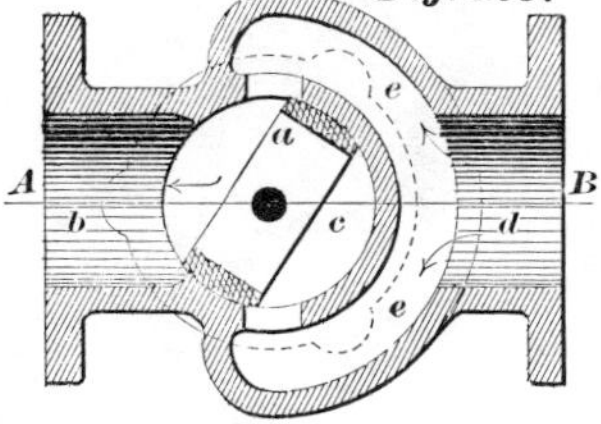

Fig. 166.

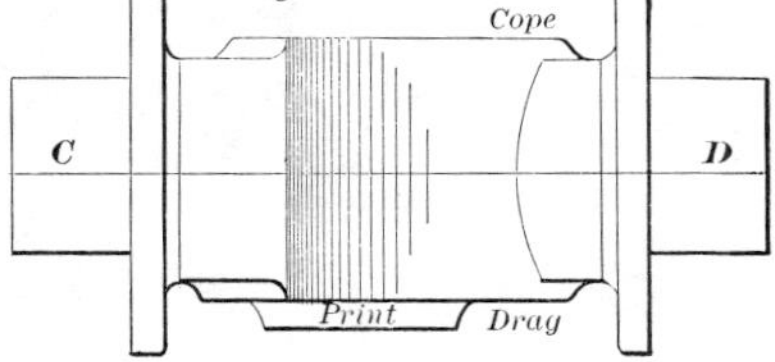

Fig. 164.

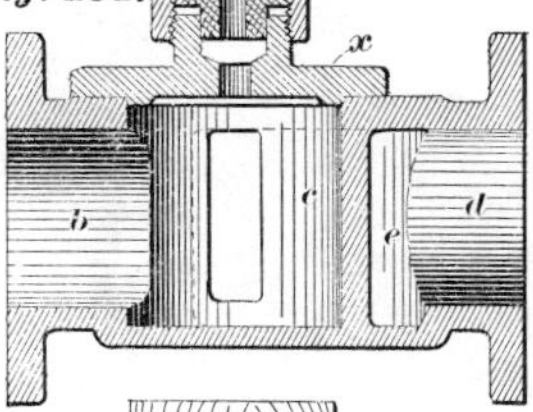

Fig. 165.

Fig. 167.

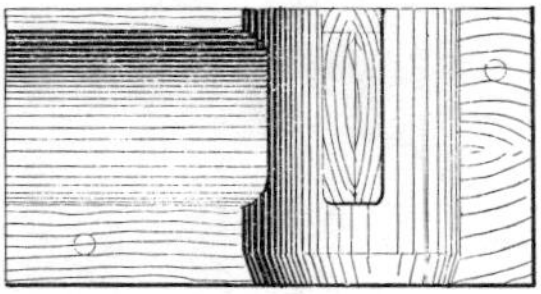

Fig. 168.

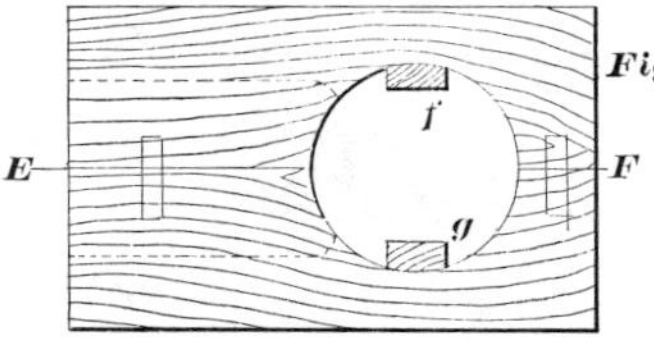

Fig. 169.

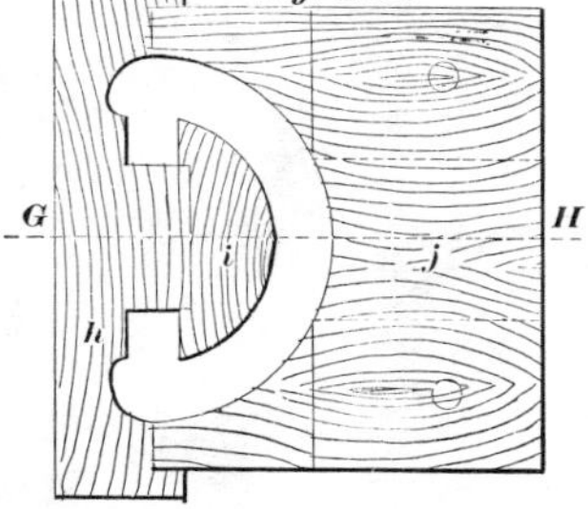

Fig. 171.

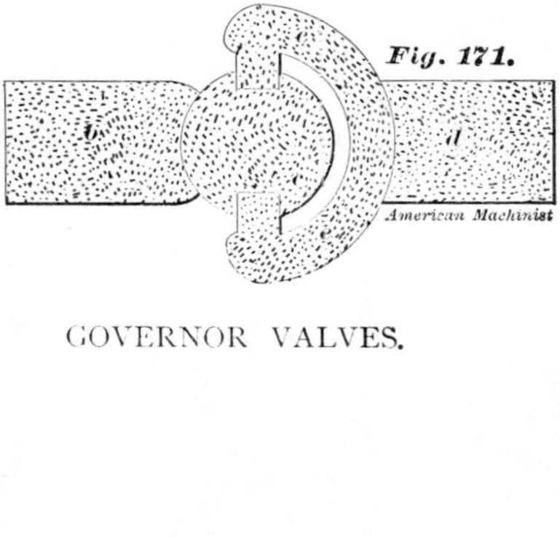

GOVERNOR VALVES.

Fig. 170.

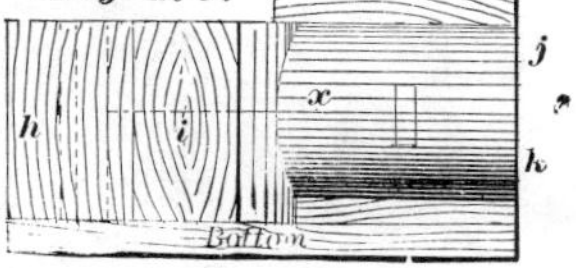

The flanges are inserted in a groove in the same manner as shown in the article on plug valves.

Figs. 167 and 168 are two views of the core-box for valve chamber *c* and outlet *b* in Figs. 163 and 164, while Figs. 169 and 170 are two views of the box for the inlet *d*, and the space *e* outside the valve chamber.

The core-box for *b c* is shown parted in Fig. 168 on line *E F*, Fig. 167 being a view of the joint on the face; *f g* are core-prints for the ports. Cores from this box need not be made in halves and pasted together in the ordinary way, as they can be made whole and dried easily otherwise.

The box in Figs. 169 and 170 is not so plain as the one just mentioned, though it is by no means a difficult box to make. By examining the two views of this box, it will be found that it is made in four parts—*h*, *i*, *j*, *k*; if the bottom be counted it will make five. Fig. 170 is a section of Fig. 169 on line *G H*—the joint line *x* in Fig. 170, shows the two pieces *j* and *k* parted and doweled together; *h* fits over the ends of *j k*, while the circular part *i* is made loose from *h*. This core also can be made whole, pasting being dispensed with.

When ramming up this core, the top side of the box will be Fig. 169; after it is rammed up, part *j* is taken off and a plate laid on top of the core, and the remaining parts of the box with the core are turned over on this plate. The bottom is then taken away, *i* drawn up, and *h* drawn away sidewise. Nothing now remains but to take off *k*, when the core will be left standing on the plate, on which it can be dried.

In Fig. 171, the two cores made from these two

boxes are shown connected, the two prints *f g* having formed a seat in *b c* for *d e* to rest in. In Fig. 167 the prints are carried flush with the top of the box, but the ports are only wanted the length indicated by arrow points. This is to alllow core *d e* to be set into the core *b c*, after which small pieces of cores are nailed or pasted into the upper parts of the slots which have not been filled by *d e*.

The dotted lines in Fig. 163 represent the bonnet *x* in Fig. 164, but as this with the other details, are plain and simple patterns to make, it will be out of place to fill valuable space with a description how to make them.

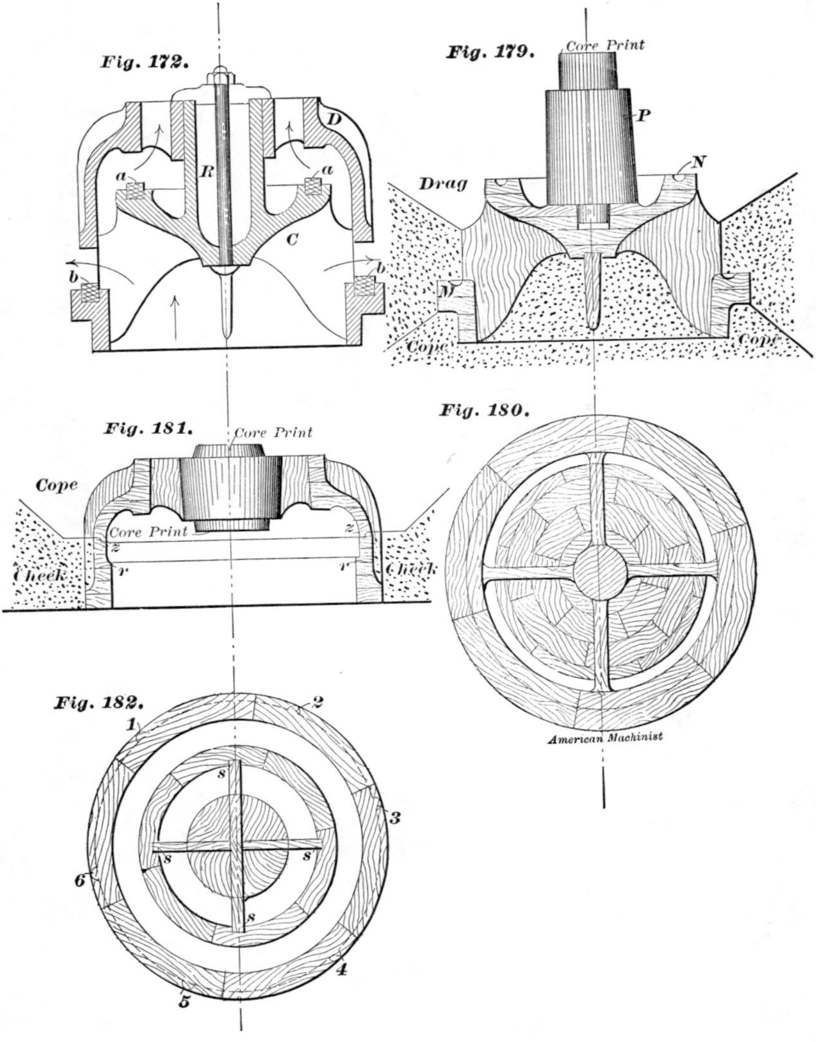

DOUBLE-BEAT VALVES.

DOUBLE-BEAT VALVES.

THE object of double-beat valves is to obtain a greater passage for the steam or water than can be had with single-beat valves of the same diameter and lift. They are sometimes called "Cornish" double-beat valves; this may be due to the fact that they are used in all large pumping engines made in the county of Cornwall, England.

There are many types of these valves, some being designed for pumps, and others for engines. One that is suitable for pumps is shown in Fig. 172, while Figs. 173 and 174 are two views of one suited for engines. The pump valve has gun metal seats inserted at *a a* and *b b*, but in Cornwall the "Bal Cap'ns" (mine managers) favor tin seats.

The valve and cage in Fig. 173 are usually made entirely of gun metal. Fig. 174 is a half section and a half plan of Fig. 173, and it is proposed to make these two patterns as shown in Figs. 175 and 176. Some exceptions may be taken to this way of making them, preference being given to solid patterns, with core-boxes for coring out the inside. This latter method would certainly give much stronger patterns, but by constructing them as shown, the amount of work in the foundry would be considerably reduced by not having any cores to make or set, except those for lightening the upper part of the valve cage at *c c*. However, should the pattern appear not strong enough for constant use (though it probably would be under ordinary circumstances), a metal pattern might be made from the wooden one, double shrinkage and finish being allowed on the original.

By referring to Fig. 175 it will be seen that the pattern for the valve cage is parted at *d d* for the convenience of moulding, it being intended to mould it in a three-part flask, allowing one part of the pattern to be drawn in an opposite direction to the other, as indicated by arrows. The seats should be cast down, to insure perfect surfaces. The four side pieces which divide the four openings should be mortised firmly into the upper ring, and two dowel pins glued and driven down through into each, as represented by dotted lines. The lower ends of these four side pieces are not secured to the lower part, but doweled together for parting.

It will be well to give the upper ring plenty of draught on the inside and outside, for, as it has to be turned and bored, it will not effect anything by so doing; also give the four side pieces plenty of taper—it will make a wonderful difference in the moulding.

Figs. 177 and 178 are two views of the core-box for coring out the upper ring, and the print to receive the segmental cores from this box is shown at *f f*. These cores will have to be secured in place by nails. In Fig. 174 it will be seen that there are small bridges at *g g*, which are formed in the core-box, print *f f* being a continuous ring. In Fig. 175 *A* represents one section of the mould; at present it is the top, but it will become the bottom, as the mould must be "rolled over"—as the

Fig. 173.

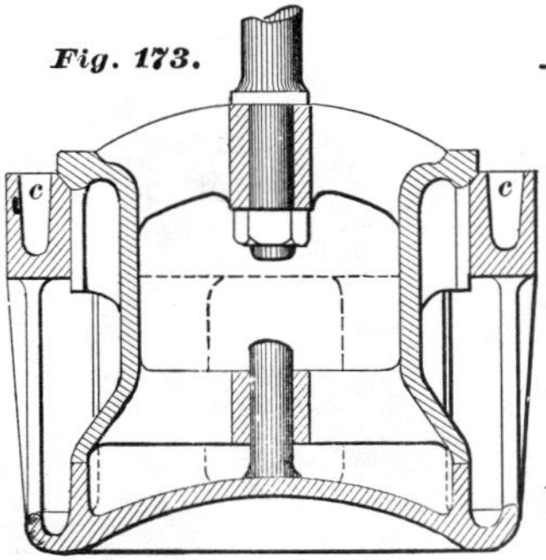

Fig. 175.

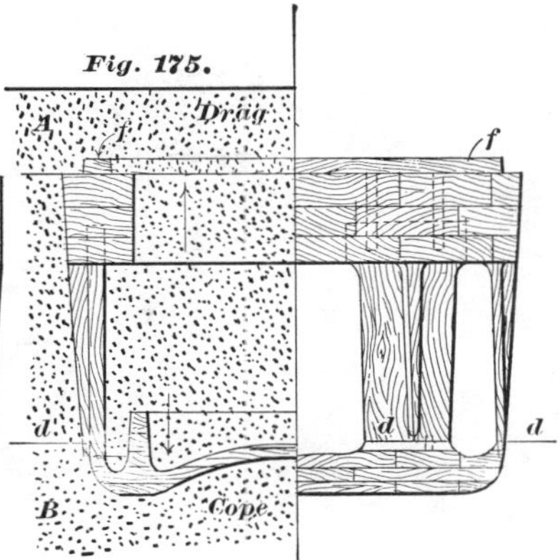

Fig. 174.

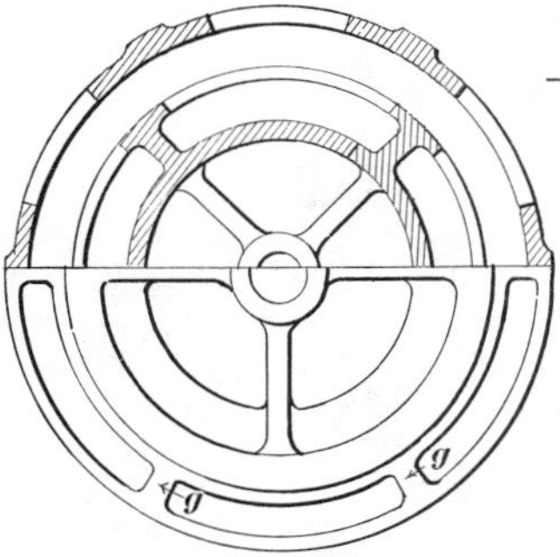

Fig. 176.

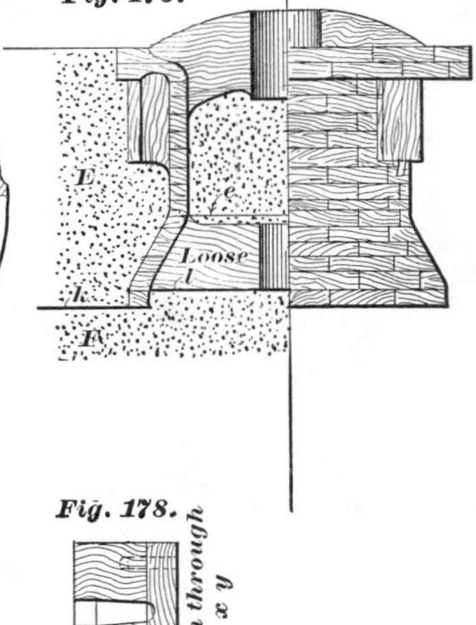

Fig. 177.

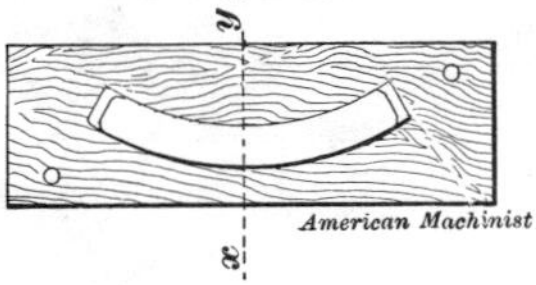

Fig. 178.

Section through x y

moulders say—in order to cast the faces of seats down. After lifting off *A*, that part of the pattern having the ring and four side pieces attached is drawn out, and the cores set in the part of mould just lifted. It is optional on the part of the moulder whether he puts on top *A* again before rolling over the mould, or rolls it without doing so; it is probably safer to put *A* on again, as it will support the centre while turning over. When this is done, section *B* of the mould is lifted off, and the remaining part of the pattern is drawn off.

The pattern for the valve (Fig. 176) is also made to leave its own core, like the cage, and is built up with segments; it is parted at *e*, one part being turned to fit into the other. The upper arms are made fast, but the lower ones should not be fastened in; the reason for this will appear later on. This valve pattern is parted at *e* for the same reason as the cage pattern in Fig. 175 is parted at *d d*, that is, for convenience of moulding, but the process of drawing these parts of the pattern differs from that in Fig. 175. The upper and lower part of the valve cage were drawn in opposite directions, as was explained, but the upper and lower part for the valve will be drawn in the same direction.

The valve is moulded and cast in the same position as shown in Fig. 176. After the cope is taken off, as represented, the pattern, as far as the joint *e*, can be drawn out, after which the cheek *E* may be lifted, allowing the lower part of the pattern to be drawn also, and leaving the loose arms behind. Before these arms can be taken out, the cheek *E* and cope must be put back in place, and the whole mould, with the bottom *F*, rolled over.

Then *F* is taken off in order to draw out the arms; the joint of *F* runs along line *k* to *l*.

These two patterns being completed, let us start in on the pump valve in Fig. 192, which is shown opened, the passage of the water being indicated by arrows.

Figs. 179 and 180 are two views of the pattern, for the stationery part *C*, and Figs. 181 and 182 for the valve *D*, which lifts up and down on the seats *a a* and *b b* in Fig. 172.

Fig. 179, like Fig 175, is cast the opposite way to that shown; this is done in order to get the valve-seats free from dirt. The ring *M* is loose, but *N* is fastened to the four wings. The centre guide *P* is also made loose, so that it can come away with the drag section of the mould. Fig. 179 represents this section of the mould taken away, but *P* is left behind for the sake of explanation. The four wings and the part *N* of the pattern are now ready to be drawn out of the sand, and this being done, the round end core that cores out the centre guide *R*, Fig. 172, should be set and tied down in the drag or lower section of the mould, and as there is no other place for the vent to be taken off, it will have to come off through the bottom, the bolt hole being drilled.

The reason for setting this core now is to save lifting off the middle after it has been rolled over onto the drag with the cope, for it is clear that if this round end core was not set before the middle was set onto the drag, that the core could not then be located.

The reader must now imagine the whole mould to be turned over in the position in which it will be cast, and with the cope taken off; he will then see that it is only required to draw the loose ring *M* out, and close the cope to complete the mould.

As to the valve *D*, a glance at the two views of the pattern in Figs. 181 and 182 will show that there is much less work about this pattern and the moulding than the one just described. Like the other style valve, it is cast in the same position as shown in Fig. 181, and the pattern is made in two parts, coming apart at *z z*. The four arms seen in Fig. 182 are half-checked together at the centre, and four corner pieces glued in, to form the hub. These arms are let into the ring at *s s*, *s s*, Fig. 182. The projection on the inside, seen in Fig. 181 at *r r*, is made so that when the casting is bored out at that place there will be a clearance for the wings in much the same way as in the counterbore of a cylinder where the piston runs over.

There are six ribs on the outside, which are numbered in Fig. 182. They are intended to prevent the valve from springing. The moulding of this pattern is similar to that in Fig. 176.

In many cases the best way to explain how a pattern should be made is to describe the manner of parting the mould. This is the plan that the writer has partly pursued in this article, the purpose being to make it clearer as to the way of constructing the patterns than otherwise could be done, though to very many readers a glance at the illustrations wlll be all that is needed in order to understand the way in which the patterns are made.

SCREW PROPELLERS.

IMAGINE a cylinder similar to that in Fig. 183, whose diameter is *A;* the curved line from *a* to *b* on the periphery represents the thread of a screw which goes exactly once around this cylinder, starting at *a* and ending at *b;* the distance from *a* to *b* is the *pitch.* Now if a right angle triangle be constructed, whose base equals the circumference of the above cylinder and the height equal to the pitch, the hypothenuse of the triangle will represent the thread from *a* to *b* unwound; that is to say, if a piece of paper is cut the same shape as the triangle and wrapped around the cylinder, the *a c* would develop the thread on its surface. To continue the thread, space off the pitch as many times as may be needed from *a*, and draw lines parallel to *a c*, as shown by dotted line *B;* then *d* will meet *a* when wound around the cylinder, thus making a continuous thread.

This simple method is the one used for marking the thread on the pattern for a worm pinion, and which some pattern makers regard as a very profound bit of wisdom that should not be allowed to become too common by "telling."

The line of the thread on the drawing, as shown in Fig. 183, is of no value for a working drawing; but the manner of developing this curve is given for the benefit of the beginner. It will be readily understood

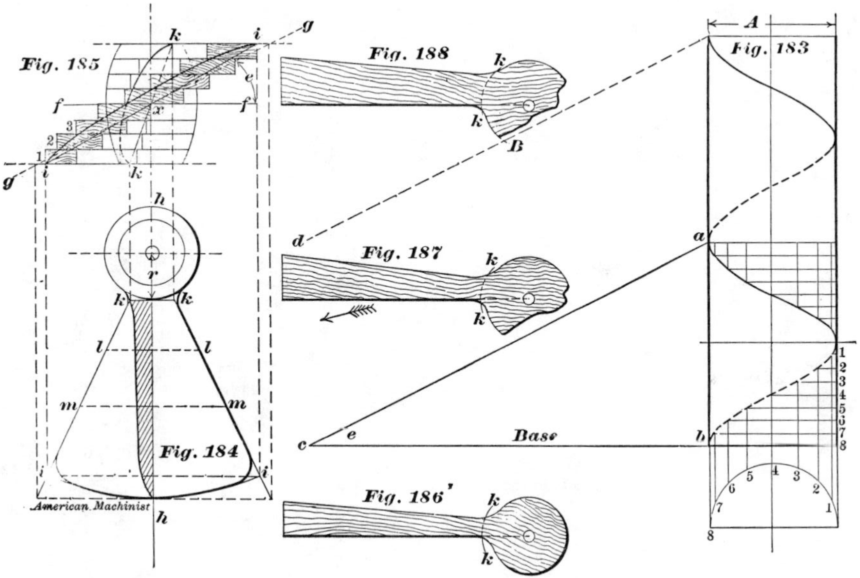
Fig. 185
Fig. 184
Fig. 188
Fig. 187
Fig. 186'
Fig. 183
A
B
Base
American Machinist

without explanation, by referring to the cross lines which are numbered.

If a three or four-blade propeller is wanted, a pattern for one blade is all that is generally needed, in which case the moulder rams up one blade at a time, moving around the pattern on a spindle.

I have seen whole patterns made for three or four-blade propellers, five and six feet in diameter, which saves much time in moulding, but unless the pattern is to be used often, which is not generally the case, it seems to be more oconomical to make but one blade, especially where pattern storage is limited. This, however, is entirely a matter that circumstance alone can decide, and be that as it may, the same principles here given for building one blade will also apply to a pattern with three or four blades.

The style of blade chosen to illustrate the manner of building the pattern is that which is used on tugboats and river steamers.

Fig. 184 is a plan of the blade, and Fig. 185 the end view. Being a true pitch it is an easy matter to get the angle at which to build the blade. First draw *f f*, which shall be the center line of the hub lengthwise; *e*, in the triangle already referred to, is the angle that the blade must be on the end. Let this angle, therefore, be made at *e* (Fig. 185) by drawing the dotted line *g g* across *f f;* then the point *x*, where these two lines intersect, is the center of the blade, as represented by *h h* in Fig. 184, and through which the face must pass.

Having located the center lines *h h*, *f f* and the angle *e*, proceed to lay out the hub, the outline and longitudinal section of the blade. The section shows the actual thickness through the center of the blade lengthwise, while the outline is not the actual but a projected one. The section across the end of the blade *i i* (Fig. 184) is seen at *i i* (Fig. 185). To get the cross-section of the blade at the hub the angle must first be made. This is done by drawing lines from *k k* (Fig. 184) until they cut the same upper and lower lines as *g g* does. Then *k k* in Fig. 185 will be the angle and section of the blade at the hub; *k k*, falling as it does on the round surface of the hub, will not appear straight, as shown, but it is drawn so in order to simplify the drawing, and as it is intended for a pattern maker's *working* drawing it will answer. The cross-section at *l l* and *m m* is omitted from both views to prevent a confusion of lines.

We now have all the lines that are needed to get the stuff out for the pattern. Care must be taken to get each course of parallel thickness, for if they are not, the building up will give trouble and cause the pitch to be wrong. A glance at the face of the blade in Fig. 185 shows we are dependent on the joints as a guide for shaping the face. For the sake of explanation, the stuff is cut as close to the line on the back of the blade as it is to the face line, but I would advise that for cutting off there be more stock allowed on the back than is represented in the drawing.

Figs. 186, 187, 188 show the shape of the first three courses, 1, 2, 3. The width of each end appears in Fig. 185, and the width at *k k* is obtained by striking a radius on each course equal to *r* in Fig. 184, and then transferring the different widths as seen at the section *k k* in Fig. 185, adding some for fillet.

Screw Propellers

Fig. 189

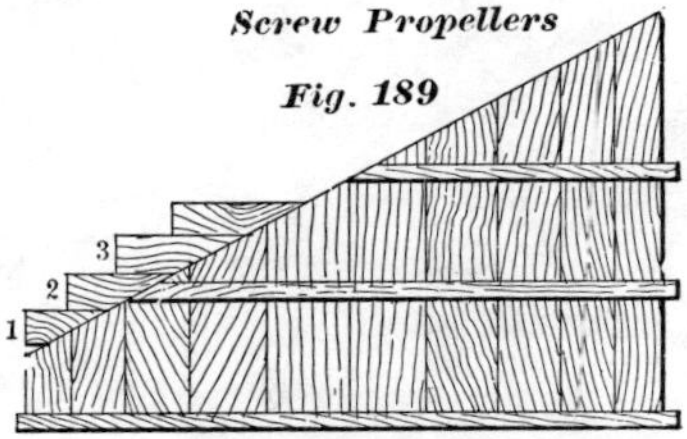

Guide and Support for end of Blade while building

In getting out the courses the grain of the wood should run as shown in the direction of arrow in Fig. 187; this will make the cutting and shaping of blade much easier than if the stuff were sawed out with the grain running straight.

A wood pin should be turned the same diameter as the spindle, on which the blade will be moved around (say 2⅞"), and a hole being bored through the hub part of each course to fit the pin, they are slipped over the pin, thus keeping all the courses concentric.

Though each course may be of even thickness, and the correct bevel laid out on the end of each piece, there is danger in the building up of the courses of their leaning over too much. To prevent this it is a good plan to get out an angle piece, as seen in Fig. 189, which is similar to that used when sweeping a propeller. This being cut to the same angle as the triangle in Fig. 183, and made to fit the same circle as the diameter of propeller, it is used as a guide and support for building up the blade. Each course for a short distance in from the end, is beveled off so as to rest on the incline of this angle piece, care being taken not to destroy the upper radial edge in doing so.

While this pattern is shown built face down, with the ends of each course resting on the angle piece, it may be built face up, and by some the latter method is preferred, because it is claimed that the pattern maker can better see to keep the face edges right when they are on the upper side, than on the lower; but this is only a matter of choice.

GEAR TEETH.

In the following pages there are a number of teeth laid out, full size, from one inch pitch to three inch, advancing by quarter-inches.

There are fourteen separate teeth in each pitch, suitable for gears having from fourteen to eight hundred teeth; they have been laid out from Prof. Robinson's Templet Odontograph and are interchangeable. The clearance allowed between the teeth is $\frac{1}{20}$ of the pitch, or in other words, the space is $\frac{55}{100}$ and the thickness of tooth $\frac{45}{100}$ of the pitch; the height of the tooth is $\frac{3}{4}$ of the pitch, and the distance from pitch line to top $\frac{33}{100}$ of the pitch. These are the proportions generally used, but the thickness of a tooth may be changed, giving more or less clearance between the teeth of a gear, as circumstance may require.

It would be impossible in a book of this kind to give the profiles of gear teeth which would serve all cases, so that I have confined myself to the system that is generally adopted, and known as the Interchangeable System, that is, all spur gears of the same pitch made under this system will run together. For special gearing and bevel gears other settings are preferred; those which I have taken are on each tooth, the setting for the flank being marked on the inside of pitch line and that for the face on the outside.

The numbers show how many different size wheels can be made with same size tooth; for instance—42 to 47 means that the same shape tooth will answer for gears which are to have from 42 to 47 teeth.

A templet can be made from the profile of any of the teeth drawn, and fastened on a rod and used in the same manner as the Odontograph itself in Fig. 96; or, if preferred, the thickness of a tooth at the top, pitch line, and bottom, may be taken and transferred on a piece of sheet zinc, and the curves struck by the aid of the Odontograph and settings given, without the usual working lines for locating the instrument. This latter way may obviate any discrepancy in the engraved teeth.

At the end of the book will be found a plate in which some of the teeth are shown in gear, together with the way they should be made.

MAKING WHEELS IN HALVES.

A VERY common practice in making large wheels is to cast them in halves by setting a cast-iron plate, or a core in the mould, cutting the hub entirely through and the rim partially, leaving sufficient metal on the outside to form a bearing surface after it is broken and bolted together.

This method of putting together large wheels of any kind cannot be considered first-class, and while it may be good enough for some wheels, for others it is not. A better way is to allow stock on one side of the centre line of each half of the wheel to permit of planing the joint; that is, to make the wheel larger in diameter one way than the other, or the joint may be planed without allowing any stock for so doing, in which case plates to go between the joint are planed to a thickness corresponding to the amount taken off both halves, and doweled on the surface of joint, but this does not give such a satisfactory job as making the two parts of the wheels a little over the centre line, as already stated.

Gear wheels which are to be in halves should never, in my opinion, be bolted together with a break joint, though this is often done, and, in making this statement, I am probably running against many who think otherwise.

It may be argued that to make a wheel in halves with a break joint is a much quicker way, and, therefore, costs

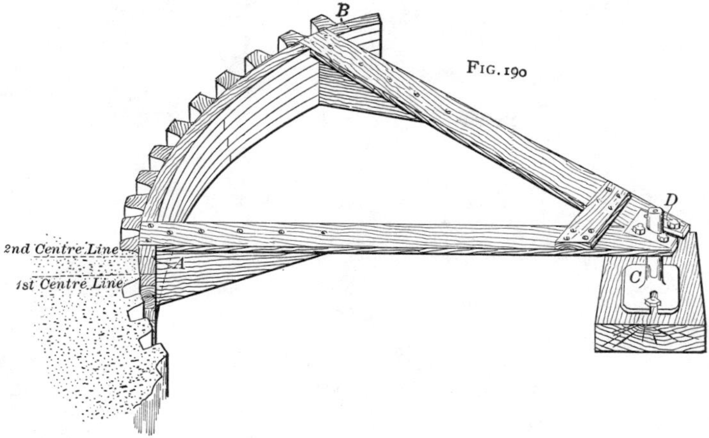

FIG. 190

FIG. 191

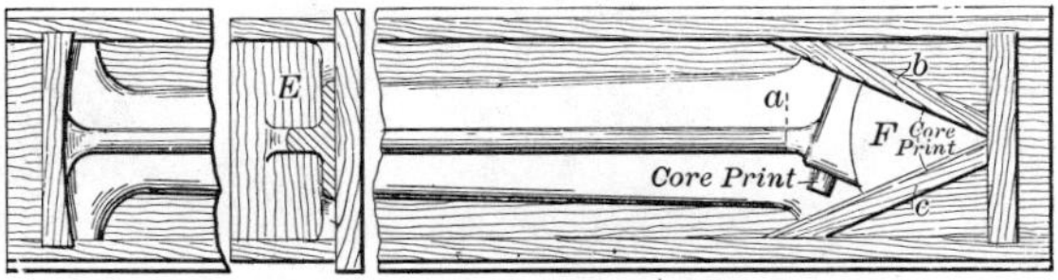

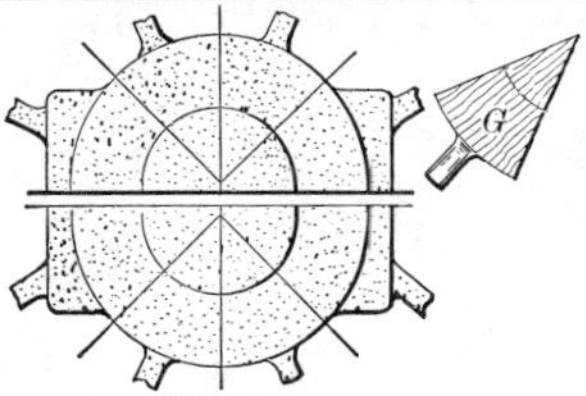

FIG. 193

FIG. 192

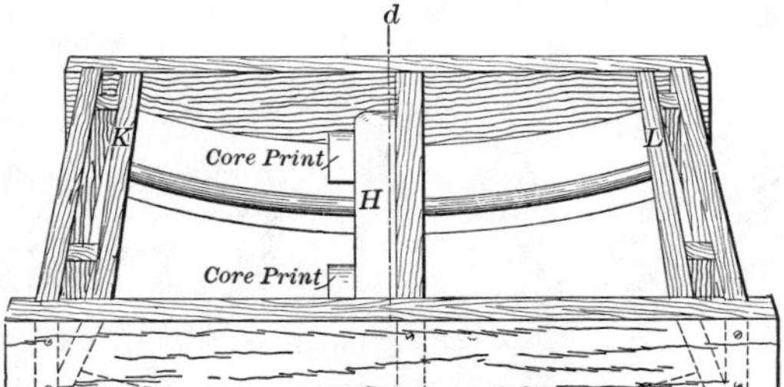

less than with a planed joint. It *is* a much quicker method, but whether it costs less in the end is a question, for sometimes a break joint in a wheel, and especially a gear wheel, gives considerable trouble after it has been running for a short time. The grain of the iron in large wheels is not usually very close, but is what foundrymen call "open,' and the coarse particles of iron in breaking are torn asunder in a way that it is almost impossible to bolt the joint together twice alike, or to have a solid bearing, because the particles do not fit perfectly into the small recesses from which they have been torn; the result is that the joint chafes and becomes loose after working awhile, and when tightened up will make the wheel run out sidewise.

To make a gear wheel in halves with a segment pattern, as represented in Fig, 190, with stock on the joint for planing, requires care and skill on the part of the moulder; with a gear moulding machine, the machine does the spacing, but in this case the moulder and pattern maker are partners, and are together responsible for the correct spacing of the teeth, and with care a gear can be made with a segment pattern that for a cast gear will be practically correct.

The pattern of the segment is built up in six courses, and in doing so every other course is cut in two. The usual way that pattern makers have in building up a large segment pattern of this kind, is to glue all the segments together, with the grain of the wood running one way; the result is that after a time the pattern springs out from the original circle it was worked to. A pattern maker would not think of building up a whole circle this way, neither should he a segment pattern that has to be kept to a certain arc like a gear segment. The thing to do after cutting every other course in two pieces, is to turn the two parts "end for end," as they say in the shop; this will keep the pattern from warping or springing out of shape if the stuff is dry—it will not be so well for working off when built as I have said it should be, but it will be right when done, which is the most important.

The radius pieces are let in flush with the top of the pattern, so that they do not interfere with the moulder when he is leveling and sleeking off the top of the mould around the teeth. It is not customary to let these pieces in, as shown; but I think it better to do so, for the reason just given. On the end is an iron plate, *D*, that fits against the spindle; this plate has four slot holes in it for adjusting the segment pattern to the correct diameter.

Referring to Fig. 190, it will be seen that a part of the segment on eaeh end is plain, having two lines, *A* and *B*, marked on the upper edge; these lines are the centres of spaces between the teeth. It is desired to make the wheel 1½″ larger in diameter one way than the other; this will allow ¼″ for planing the joint, and for splitting cores 1″ thick. In this case it is best to use cores for splitting, in preference to cast iron plates, as the iron plates would chill the surface to be planed, making it hard, and thus causing the planer hand trouble. Suppose this gear to have ninety-six teeth. After forty-eight or one-half of the teeth have been formed, two centre lines, 1½″ apart, should be made on the top of the mould, as shown, representing the joint of casting after it has been planed. Now by moving the spindle 1½″ at right angles to these lines, and setting the line *A*, marked on the pattern, to the second centre line, it will give the necessary space between the teeth next the joint, allowing for 1″ splitting core, and ¼″ for planing.

It will be noticed that on the end of the spindle is a square base *C*, into which two holes are cast for bolting it to a timber. These holes are slotted, enabling the moulder to move the center as described. The spindle is cored out like a pipe, and if bored a little tapering may be used as a socket for an ordinary spindle to work in. Such a spindle will be found a very handy, though simple, rig for making wheels in halves, when joints have to be planed.

I have said that the hub is cut entirely through; this should always be done, whether the halves are to be planed or not, as it relieves the shrinking strain in the arms and rim.

The inside of the rim and the arms of the wheel are formed with cores. Boxes from which these cores are made as shown in Figs. 191 and 192. A half section of the arm, which is a cross shape section, is seen at *E*, and as in the case of oval section arms, a core box for one-half of the arm is all that is required. There are three loose pieces fitted in the end of the box to form the hub, one marked *F*, another the opposite hand to *F*, and the plain piece marked *G* shown outside the arm box. Each of these pieces matches the rib of arm at *a*. The two short sides *b* and *c* of the box are set ¼″ back from the radial line, or where the cores join, and two loose pieces

¼″ thick fitted in, to bring the sides to the radial line. The object of this is to give ¼″ stock on the hub, for planing.

When *F* is used, the ¼″ piece on side *b* is left out, while the one on side *c* remains in, and when the opposite part to *F* is used, then the ¼″ piece on side *c* is left out, and the one for side *b* is put in. When *G* is used, then the ¼″ pieces go in on both sides.

There being eight arms to this gear wheel, there will be four cores each when using *F*, and the piece which is the opposite hand to *F*, and eight with *G*, making sixteen for the bottom and top.

When these arm cores are located inside the mould whose diameter, it will be remembered, is made 1½″ longer one way than the other, the cores at the hub will be 1″ apart, as seen in Fig. 193; this space will receive the splitting core after the two half-round cores for the bore have been set in place.

The core box, from which the cores are made to form the inside of the rim between the arms, is shown in Fig. 192. *H* is the lug for bolting the wheel together at the rim, and is taken out after making four short cores. These cores, of course, are located in the mould, so that the lugs correspond with split in the hub, and when set will leave a space to receive the splitting core in the same way as the arm cores at the center in Fig. 193. The lug *H* being set ¼″ past the center line *d* shows the stock that is allowed for planing the joint.

The two ends *L* and *K* of this box run under, making a "back draft," as represented by dotted lines on the side; they are made thus, so that the ends of cores may fit against the sides of the arm cores. These ends being "back draft" are, therefore, made loose, in order that the cores may leave the box without taking it apart—the ends come out with the core.

That part of the rim at the lugs is not cut into by the splitting cores; the rim is, therefore, the full thickness at that point. This is done to prevent the wheel from breaking at the joint, and the ends from springing away from the center lines while shrinking, and that the casting might remain intact for testing and laying out the two center lines.

Having once made two gears 10 ft. 2¼″ diameter, 4″ pitch, 12″ face, in the manner described here, I had some misgiving that the rims might spring after they had been cut in two, causing the teeth next the joint to be out of place, but they did not, and after they were planed and bolted together the spaces between the teeth at the joint did not vary from the others. The wheels, also, were practically true, thus showing that with extra care on the part of moulder and pattern maker, a good job may be done in making gears this way and without the gear moulding machine. I do not mean by this to depreciate in any way gear moulding machines, but then they are not on hand in most shops when needed.

Table of the Diameter of Wheels at the Pitch Circle, from 11 to 300 Teeth.

Number of Teeth.	Pitch of the Teeth.									
	inch. 1¾.	inch. 1⅞.	inches. 2	inches. 2⅛.	inches. 2¼.	inches. 2½.	inches. 2¾.	inches. 3.	inches. 3¼.	inches. 3½.
11	0 6¼	0 6⅝	0 7	0 7½	0 7⅞	0 8¾	0 9¾	0 10⅝	0 11⅜	1 0¼
12	0 6¾	0 7⅛	0 7⅝	0 8⅛	0 8⅝	0 9⅝	0 10⅝	0 11½	1 0⅜	1 1⅜
13	0 7⅜	0 7⅞	0 8⅜	0 8⅞	0 9⅜	0 10⅜	0 11½	1 0½	1 1⅜	1 2½
14	0 7⅞	0 8½	0 9	0 9½	0 10	0 11¼	1 0⅜	1 1½	1 2½	1 3⅝
15	0 8½	0 9	0 9⅝	0 10¼	0 10¾	1 0	1 1¼	1 2⅜	1 3½	1 4¾
16	0 9	0 9⅝	0 10¼	0 10⅞	0 11½	1 0¾	1 2	1 3⅜	1 4½	1 5⅞
17	0 9⅝	0 10¼	0 10⅞	0 11½	1 0¼	1 1½	1 2⅞	1 4¼	1 5½	1 6⅞
18	0 10	0 10¾	0 11½	1 0¼	1 0⅞	1 2⅜	1 3¾	1 5¼	1 6⅝	1 8
19	0 10⅝	0 11⅜	1 0⅛	1 0⅞	1 1⅝	1 3⅛	1 4⅝	1 6¼	1 7⅝	1 9⅛
20	0 11¼	1 0	1 0¾	1 1½	1 2⅜	1 4	1 5½	1 7⅛	1 8¾	1 10¼
21	0 11¾	1 0½	1 1½	1 2¼	1 3	1 4¾	1 6⅜	1 8⅛	1 9¾	1 11⅜
22	1 0⅜	1 1⅛	1 2	1 2⅞	1 3¾	1 5½	1 7¼	1 9	1 10¾	2 0½
23	1 0⅞	1 1¾	1 2⅝	1 3½	1 4½	1 6⅜	1 8	1 10	1 11¾	2 1⅝
24	1 1½	1 2⅜	1 3⅜	1 4¼	1 5¼	1 7⅛	1 9	1 10⅞	2 0⅞	2 2¾
25	1 2	1 2⅞	1 3⅞	1 4⅞	1 6	1 8	1 9⅞	1 11⅞	2 1⅞	2 3⅞
26	1 2½	1 3½	1 4½	1 5½	1 6⅝	1 8¾	1 10¾	2 0⅞	2 2⅞	2 4⅞
27	1 3	1 4⅛	1 5¼	1 6¼	1 7⅜	1 9½	1 11⅝	2 1¾	2 3⅞	2 6⅛
28	1 3⅝	1 4⅝	1 5¾	1 6⅞	1 8	1 10¼	2 0½	2 2¾	2 4⅞	2 7¼
29	1 4⅛	1 5⅜	1 6½	1 7⅝	1 8¾	1 11⅛	2 1⅜	2 3¾	2 6	2 8⅜
30	1 4½	1 6	1 7⅛	1 8¼	1 9½	2 0	2 2¼	2 4⅝	2 7	2 9⅜
31	1 5⅜	1 6½	1 7¾	1 9	1 10¼	2 0¾	2 3⅛	2 5⅝	2 8	2 10½
32	1 5⅞	1 7⅛	1 8⅜	1 9⅝	1 11	2 1½	2 4	2 6½	2 9⅛	2 11⅝
33	1 6½	1 7¾	1 9	1 10⅜	1 11⅝	2 2¼	2 4⅞	2 7½	2 10⅛	3 0¾
34	1 7	1 8⅜	1 9⅝	1 11	2 0⅜	2 3	2 5¾	2 8½	2 11⅛	3 1⅞
35	1 7½	1 9	1 10⅛	1 11¾	2 1	2 3⅞	2 6⅝	2 9½	3 0¼	3 3
36	1 8	1 9½	1 10⅞	2 0⅜	2 2	2 4⅝	2 7½	2 10⅜	3 1¼	3 4⅛
37	1 8⅝	1 10	1 11½	2 1	2 2½	2 5½	2 8⅜	2 11⅜	3 2¼	3 5¼
38	1 9¼	1 10¾	2 0¼	2 1¾	2 3¼	2 6¼	2 9¼	3 0¼	3 3⅜	3 6⅜
39	1 9¾	1 11⅜	2 0⅞	2 2⅜	2 4	2 7	2 10⅛	3 1¼	3 4⅜	3 7⅝
40	1 10⅜	1 11⅞	2 1½	2 3	2 4⅝	2 7⅞	2 10¾	3 2¼	3 5⅜	3 8½
41	1 10⅞	2 0½	2 2⅛	2 3¾	2 5⅜	2 8⅝	2 11⅞	3 3⅛	3 6⅜	3 9⅝
42	1 11½	2 1	2 2¾	2 4½	2 6	2 9⅜	3 0¾	3 4⅛	3 7⅜	3 10¾
43	2 0	2 1⅝	2 3⅜	2 5	2 6¾	2 10¼	3 1⅝	3 5	3 8⅜	4 0
44	2 0½	2 2¼	2 4	2 5¾	2 7½	2 11	3 2½	3 6	3 9½	4 1
45	2 1	2 2⅞	2 4⅝	2 6½	2 8¼	2 11¾	3 3⅜	3 7	3 10½	4 2¼
46	2 1⅝	2 3½	2 5¼	2 7⅛	2 9	3 0⅝	3 4¼	3 7⅞	3 11½	4 3¼
47	2 1⅞	2 4	2 6	2 7⅞	2 9⅝	3 1½	3 5⅛	3 8⅞	4 0⅝	4 4⅜
48	2 2¾	2 4⅝	2 6½	2 8½	2 10⅜	3 2¼	3 6	3 9⅞	4 1⅝	4 5½
49	2 3⅜	2 5¼	2 7¼	2 9⅛	2 11	3 3	3 6⅞	3 10⅞	4 2⅝	4 6½
50	2 3⅞	2 5⅞	2 7⅞	2 9¾	2 11¾	3 3¾	3 7¾	3 11¾	4 3¾	4 7¾
51	2 4½	2 6½	2 8½	2 10½	3 0½	3 4½	3 8⅝	4 0¾	4 4¾	4 8⅞
52	2 4⅞	2 7⅛	2 9⅛	2 11⅛	3 1¼	3 5⅜	3 9½	4 1⅝	4 5¾	4 10
53	2 5⅜	2 7⅝	2 9¾	2 11⅞	3 2	3 6¼	3 10½	4 2⅝	4 6⅞	4 11
54	2 6	2 8¼	2 10⅜	3 0½	3 2⅝	3 7	3 11	4 3½	4 7⅞	5 0⅛
55	2 6⅝	2 8⅞	2 11	3 1¼	3 3½	3 7¾	4 0⅛	4 4½	4 8⅞	5 1¼
56	2 7⅛	2 9⅜	2 11⅝	3 1⅞	3 4⅛	3 8¼	4 1	4 5½	4 9⅞	5 2⅜
57	2 7¾	2 10	3 0¼	3 2½	3 4⅞	3 9⅜	4 1⅞	4 6⅜	4 10⅞	5 3½

Table of the Diameter of Wheels at the Pitch Circle—Continued.

Number of Teeth.	Pitch of the Teeth. inch. 1¾.	inch. 1⅞.	inches. 2.	inches. 2⅛.	inches. 2¼.	inches. 2½.	inches. 2¾.	inches. 3.	inches. 3¼.	inches. 3½.
58	2 8¼	2 10⅝	3 0⅞	3 3¼	3 5½	3 10⅛	4 2¾	4 7⅜	5 0	5 4⅝
59	2 8⅞	2 11¼	3 1½	3 4	3 6¼	3 11⅛	4 3⅝	4 8⅜	5 1	5 5¾
60	2 9⅜	2 11¾	3 2¼	3 4⅝	3 7	3 11¾	4 4½	4 9¼	5 2	5 6⅞
61	2 10	3 0⅜	3 2⅞	3 5¼	3 7¾	4 0½	4 5¾	4 10¼	5 3⅛	5 8
62	2 10½	3 1	3 3½	3 6	3 8½	4 1⅜	4 6¼	4 11¼	5 4⅛	5 9
63	2 11	3 1⅝	3 4⅛	3 6⅝	3 9⅛	4 2⅛	4 7⅛	5 0⅛	5 5⅛	5 10⅛
64	2 11⅝	3 2¼	3 4¾	3 7¼	3 9⅞	4 3	4 8	5 1⅛	5 6¼	5 11⅜
65	3 0¼	3 2⅞	3 5⅜	3 8	3 10½	4 3¾	4 8⅞	5 2	5 7¼	6 0⅜
66	3 0⅞	3 3⅜	3 6	3 8⅝	3 11¼	4 4½	4 9¾	5 3	5 8¼	6 1½
67	3 1⅜	3 4	3 6½	3 9⅜	4 0	4 5⅜	4 10⅝	5 4	5 9⅜	6 2⅝
68	3 1¾	3 4⅝	3 7¼	3 10	4 0¾	4 6⅛	4 11½	5 5	5 10⅜	6 3¾
69	3 2⅜	3 5⅛	3 7⅞	3 10⅝	4 1½	4 7	5 0⅜	5 6	5 11⅜	6 4⅞
70	3 3	3 5¾	3 8½	3 11⅜	4 2⅛	4 7¾	5 1¼	5 6⅞	6 0½	6 6
71	3 3½	3 6⅜	3 9¼	4 0	4 2⅞	4 8½	5 2⅛	5 7¾	6 1½	6 7
72	3 4⅛	3 6¾	3 9⅞	4 0¾	4 3½	4 9¼	5 3	5 8¾	6 2½	6 8¼
73	3 4⅝	3 7½	3 10½	4 1⅜	4 4¼	4 10	5 3⅞	5 9¾	6 3½	6 9⅜
74	3 5¼	3 7⅞	3 11⅛	4 2	4 5	4 10⅞	5 4¾	5 10⅝	6 4½	6 10⅝
75	3 5¾	3 8¾	3 11¾	4 2¾	4 5¾	4 11¾	5 5⅜	5 11⅝	6 5½	6 11½
76	3 6⅜	3 9⅜	4 0⅜	4 3½	4 6½	5 0½	5 6½	6 0½	6 6⅝	7 0⅝
77	3 6⅞	3 9⅞	4 1	4 4	4 7⅛	5 1¼	5 7⅜	6 1½	6 7⅝	7 1¼
78	3 7½	3 10½	4 1⅝	4 4¾	4 7⅞	5 2	5 8¼	6 2½	6 8⅝	7 2⅞
79	3 8	3 11⅛	4 2¼	4 5½	4 8½	5 2⅞	5 9⅛	6 3½	6 9¾	7 4
80	3 8½	3 11¾	4 3	4 6⅛	4 9¼	5 3⅝	5 10	6 4⅜	6 10¾	7 5⅛
81	3 9⅛	4 0⅜	4 3½	4 6⅞	4 10	5 4½	5 10⅞	6 5⅜	6 11⅞	7 6¼
82	3 9⅝	4 0⅞	4 4¼	4 7½	4 10¾	5 5¼	5 11¾	6 6⅜	7 0⅞	7 7⅜
83	3 10¼	4 1½	4 4⅞	4 8⅛	4 11½	5 6	6 0⅝	6 7¼	7 1⅞	7 8⅝
84	3 10¾	4 2⅛	4 5½	4 8⅞	5 0⅛	5 6⅞	6 1½	6 8⅛	7 2⅞	7 9½
85	3 11¼	4 2¾	4 6⅛	4 9½	5 0⅞	5 7⅝	6 2⅜	6 9⅛	7 3⅞	7 10⅝
86	3 11⅞	4 3¼	4 6¾	4 10⅛	5 1⅝	5 8½	6 3¼	6 10½	7 5	7 11¾
87	4 0½	4 3⅞	4 7⅜	4 10⅞	5 2¼	5 9¼	6 4⅛	6 11	7 6	8 0⅞
88	4 1	4 4½	4 8	4 11½	5 3	5 10	6 5	7 0	7 7	8 2
89	4 1½	4 5⅛	4 8⅝	5 0¼	5 3¼	5 10¾	6 5⅞	7 1	7 8	8 3⅛
90	4 2⅛	4 5¾	4 9¼	5 0⅞	5 4¼	5 11⅝	6 6¾	7 2	7 9⅛	8 4¼
91	4 2¾	4 6¼	4 9⅞	5 1½	5 5⅛	6 0⅜	6 7⅝	7 2⅞	7 10⅛	8 5⅜
92	4 3¼	4 7	4 10½	5 2⅛	5 5⅞	6 1	6 8½	7 3⅞	7 11⅛	8 6½
93	4 3⅞	4 7½	4 11¼	5 2⅞	5 6⅝	6 2	6 9⅜	7 4⅞	8 0¼	8 7⅝
94	4 4⅜	4 8⅛	4 11¾	5 3½	5 7⅜	6 2¾	6 10¼	7 5¾	8 1¼	8 8¾
95	4 4⅞	4 8¾	5 0½	5 4¼	5 8	6 3½	6 11⅛	7 6¾	8 2¼	8 9⅞
96	4 5½	4 9⅜	5 1⅛	5 5	5 8¾	6 4⅜	7 0	7 7⅝	8 3⅜	8 10⅞
97	4 6	4 10	5 1¾	5 5⅝	5 9½	6 5¼	7 0⅞	7 8⅝	8 4⅜	9 0
98	4 6½	4 10½	5 2⅜	5 6¼	5 10⅛	6 6	7 1¾	7 9½	8 5⅜	9 1⅛
99	4 7⅛	4 11	5 3	5 7	5 11	6 6¾	7 2⅝	7 10½	8 6½	9 2¼
100	4 7¾	4 11⅝	5 3⅝	5 7⅝	5 11⅝	6 7½	7 3⅝	7 11½	8 7½	9 3⅜
101	4 8¼	5 0¼	5 4¼	5 8¼	6 0¼	6 8⅜	7 4⅜	8 0½	8 8½	9 4½
102	4 8⅞	5 1	5 5	5 9	6 1	6 9⅛	7 5¼	8 1⅜	8 9½	9 5⅝
103	4 9⅜	5 1¼	5 5½	5 9⅝	6 1½	6 10	7 6⅛	8 2¼	8 10½	9 6¾
104	4 10	5 1¾	5 6¼	5 10¼	6 2¼	6 10¾	7 7	8 3¼	8 11½	9 7⅞

Table of the Diameter of Wheels at the Pitch Circle—Continued.

Number of Teeth.	Pitch of the Teeth.									
	inch. 1¾.	inch. 1⅞.	inches. 2.	inches. 2⅛.	inches. 2¼.	inches. 2½.	inches. 2¾.	inches. 3.	inches. 3¼.	inches. 3½.
105	4 10½	5 2⅜	5 6¾	5 11	6 3	6 11½	7 7⅞	8 4¼	9 0⅝	9 8⅞
106	4 11	5 2⅞	5 7½	5 11⅝	6 3⅝	7 0¼	7 8¾	8 5⅛	9 1⅝	9 10
107	4 11½	5 3½	5 8½	6 0⅜	6 4¼	7 1⅛	7 9⅝	8 6⅛	9 2⅝	9 11¼
108	5 0⅛	5 4⅛	5 8¾	6 1	6 5	7 2	7 10½	8 7⅛	9 3¾	10 0⅜
109	5 0¾	5 4¾	5 9⅜	6 1¾	6 5¾	7 2¾	7 11⅜	8 8	9 4¾	10 1⅜
110	5 1¼	5 5¼	5 10	6 2⅜	6 6½	7 3½	8 0¼	8 9	9 5¾	10 2⅜
111	5 1⅞	5 5⅞	5 10⅝	6 3	6 7⅛	7 4¼	8 1⅛	8 10	9 6⅞	10 3⅝
112	5 2⅜	5 6½	5 11¼	6 3¾	6 8	7 5⅛	8 2	8 10⅞	9 7⅞	10 4¾
113	5 3	5 7	6 0	6 4⅜	6 8⅝	7 6	8 3	8 11⅞	9 8⅞	10 6
114	5 3½	5 7⅝	6 0½	6 5⅛	6 9¼	7 6¾	8 3¾	9 0¾	9 9⅞	10 7
115	5 4	5 8¼	6 1¼	6 5¾	6 10	7 7½	8 4⅝	9 1¾	9 10⅞	10 8⅛
116	5 4⅝	5 8⅞	6 1¾	6 6½	6 10¾	7 8¼	8 5½	9 2¾	10 0	10 9¼
117	5 5⅛	5 9½	6 2½	6 7⅛	6 11½	7 9⅛	8 6⅜	9 3⅝	10 1	10 10⅜
118	5 5¾	5 10	6 3⅛	6 7¾	7 0⅛	7 10	8 7¼	9 4⅝	10 2	10 11⅝
119	5 6¼	5 10⅝	6 3¾	6 8½	7 1	7 10¾	8 8⅛	9 5½	10 3⅛	11 0½
120	5 6¾	5 11¼	6 4⅜	6 9⅛	7 1⅝	7 11½	8 9	9 6½	10 4⅛	11 1⅝
121	5 7⅜	5 11⅞	6 5	6 9¾	7 2¼	8 0¼	8 9⅞	9 7½	10 5⅛	11 2¾
122	5 8	6 0½	6 5⅝	6 10½	7 3	8 1	8 10¾	9 8½	10 6¼	11 3⅞
123	5 8½	6 1	6 6¼	6 11⅛	7 3¾	8 1⅞	8 11⅝	9 9⅜	10 7¼	11 5
124	5 9	6 1⅝	6 7	6 11⅞	7 4½	8 2⅝	9 0½	9 10⅜	10 8¼	11 6⅛
125	5 9⅝	6 2¼	6 7½	7 0½	7 5⅛	8 3½	9 1⅜	9 11¼	10 9⅜	11 7¼
126	5 10⅛	6 2¾	6 8¼	7 1¼	7 6	8 4¼	9 2¼	10 0¼	10 10⅜	11 8⅜
127	5 10¾	6 3⅜	6 8¾	7 2	7 6⅝	8 5	9 3⅛	10 1¼	10 11⅜	11 9½
128	5 11¼	6 4	6 9½	7 2½	7 7¼	8 5⅞	9 4	10 2¼	11 0½	11 10½
129	5 11⅞	6 4½	6 10⅛	7 3¼	7 8	8 6⅝	9 4⅞	10 3⅛	11 1½	11 11¾
130	6 0⅜	6 5⅛	6 10¾	7 4	7 8¾	8 7½	9 5¾	10 4⅛	11 2½	12 0⅞
131	6 1	6 5¾	6 11⅜	7 4½	7 9½	8 8¼	9 6⅝	10 5	11 3½	12 2
132	6 1½	6 6⅜	7 0	7 5⅛	7 10⅛	8 9	9 7½	10 6	11 4⅝	12 3
133	6 2	6 7	7 0⅝	7 6	7 10⅞	8 9¾	9 8⅜	10 7	11 5⅝	12 4⅛
134	6 2⅝	6 7½	7 1¼	7 6⅝	7 11½	8 10⅝	9 9¼	10 8	11 6⅝	12 5¼
135	6 3⅛	6 8½	7 2	7 7¼	8 0¼	8 11⅜	9 10⅛	10 8⅞	11 7⅝	12 6¾
136	6 3¾	6 8¾	7 2½	7 8	8 1	9 0¼	9 11	10 9¾	11 8⅝	12 7½
137	6 4¼	6 9¼	7 3¼	7 8⅝	8 1¾	9 1	10 0	10 10¾	11 9¾	12 8⅝
138	6 4¾	6 10	7 3¾	7 9¼	8 2½	9 1¾	10 0⅞	10 11¾	11 10¾	12 9¾
139	6 5½	6 10½	7 4½	7 10	8 3⅛	9 2⅝	10 1⅝	11 0⅝	11 11¾	12 10⅞
140	6 6	6 11⅛	7 5⅛	7 10⅝	8 3⅞	9 3⅜	10 2½	11 1⅝	12 0⅞	13 0
141	6 6½	6 11¾	7 5¾	7 11⅜	8 4½	9 4¼	10 3⅜	11 2½	12 1⅞	13 1
142	6 7	7 0¼	7 6¾	8 0	8 5¼	9 5	10 4¼	11 3½	12 2⅞	13 2⅛
143	6 7⅝	7 0⅞	7 7	8 0¾	8 6	9 5¾	10 5⅛	11 4½	12 3⅞	13 3⅜
144	6 8¼	7 1½	7 7⅝	8 1⅜	8 6¾	9 6	10 6	11 5½	12 4⅞	13 4⅜
145	6 8¾	7 2	7 8¼	8 2	8 7½	9 7⅜	10 6⅞	11 6½	12 6	13 5½
146	6 9¼	7 2⅝	7 9	8 2¾	8 8½	9 8⅛	10 7¾	11 7⅜	12 7	13 6⅝
147	6 9⅞	7 3¼	7 9½	8 3⅜	8 8⅞	9 9	10 8⅝	11 8¼	12 8	13 7¾
148	6 10½	7 3⅞	7 10¼	8 4⅛	8 9½	9 9¾	10 9½	11 9¼	12 9⅛	13 8⅞
149	6 11	7 4½	7 10¾	8 4⅞	8 10¼	9 10½	10 10⅜	11 10¼	12 10⅛	13 10
150	6 11½	7 5	7 11½	8 5½	8 11	9 11⅜	10 11¼	11 11¼	12 11⅛	13 11⅛

Number of Teeth.	Pitch of the Teeth.					
	inches. 2¼.	inches. 2½.	inches. 2¾.	inches. 3.	inches. 3¼.	inches. 3½.
151	9 0⅛	10 0⅛	11 0⅛	12 0⅛	13 0⅛	14 0¼
152	9 0⅞	10 0⅞	11 1	12 1⅛	13 1¼	14 1⅜
153	9 1⅝	10 1¾	11 1⅞	12 2	13 2¼	14 2⅜
154	9 2¼	10 2½	11 2⅞	12 3	13 3⅜	14 3½
155	9 3	10 3⅜	11 3⅝	12 4	13 4⅜	14 4⅝
156	9 3⅝	10 4⅛	11 4½	12 4⅞	13 5⅜	14 5¾
157	9 4⅜	10 4⅞	11 5⅝	12 5⅞	13 6⅜	14 6⅞
158	9 5⅛	10 5⅝	11 6⅜	12 6⅞	13 7⅜	14 8
159	9 5⅞	10 6½	11 7⅛	12 7⅞	13 8⅜	14 9⅛
160	9 6½	10 7⅜	11 8	12 8¾	13 9½	14 10¼
161	9 7⅜	10 8⅛	11 8⅞	12 9⅝	13 10½	14 11⅜
162	9 8	10 8⅞	11 9¾	12 10⅝	13 11½	15 0⅜
163	9 8¾	10 9⅝	11 10⅝	12 11⅝	14 0¾	15 1½
164	9 9⅜	10 10½	11 11½	13 0⅝	14 1⅝	15 2⅝
165	9 10⅛	10 11⅜	12 0¾	13 1½	14 2⅝	15 3¾
166	9 10⅞	11 0	12 1¼	13 2½	14 3¾	15 4⅞
167	9 11⅝	11 0⅞	12 2⅛	13 3⅜	14 4⅜	15 6
168	10 0⅜	11 1⅝	12 3	13 4⅜	14 5¾	15 7⅛
169	10 1	11 2⅜	12 3⅞	13 5⅜	14 6⅝	15 8¼
170	10 1¾	11 3¼	12 4¾	13 6⅜	14 7⅝	15 9⅝
171	10 2⅜	11 4	12 5⅝	13 7¼	14 8⅞	15 10½
172	10 3⅛	11 4⅞	12 6½	13 8⅛	14 9⅞	15 11⅝
173	10 3⅞	11 5¾	12 7⅜	13 9⅛	14 10⅞	16 0⅝
174	10 4½	11 6⅜	12 8¼	13 10⅛	15 0	16 1¾
175	10 5⅜	11 7¼	12 9⅛	13 11⅛	15 1⅜	16 2⅞
176	10 6	11 8	12 10	14 0	15 2⅝	16 4
177	10 6¾	11 8⅞	12 10⅞	14 1	15 3⅛	16 5⅛
178	10 7⅝	11 9⅝	12 11⅞	14 1⅞	15 4¼	16 6¾
179	10 8⅛	11 10⅜	13 0⅝	14 2⅞	15 5⅛	16 7⅜
180	10 8⅞	11 11¼	13 1½	14 3⅞	15 6¼	16 8½
181	10 9⅝	12 0	13 2⅝	14 4⅞	15 7¼	16 9⅝
182	10 10⅜	12 0⅞	13 3⅜	14 5¾	15 8¼	16 10¾
183	10 11	12 1⅝	13 4⅛	14 6¾	15 9⅜	16 11⅞
184	10 11¾	12 2⅜	13 5	14 7⅝	15 10⅜	17 0⅞
185	11 0⅜	12 3⅛	13 5⅞	14 8⅝	15 11⅜	17 2⅛
186	11 1⅛	12 4	13 6⅞	14 9⅝	16 0⅜	17 3¼
187	11 1⅞	12 4⅞	13 7⅝	14 10½	16 1⅜	17 4⅜
188	11 2⅝	12 5½	13 8½	14 11½	16 2⅜	17 5⅜
189	11 3⅜	12 6⅜	13 9⅜	15 0⅜	16 3½	17 6½
190	11 4	12 7⅛	13 10⅜	15 1⅜	16 4½	17 7⅝
191	11 4¾	12 7⅞	13 11⅛	15 2⅜	16 5½	17 8¾
192	11 5½	12 8⅜	14 0	15 3⅜	16 6⅝	17 9⅞
193	11 6⅛	12 9½	14 0⅞	15 4¼	16 7⅝	17 11
194	11 6⅞	12 10⅜	14 1⅞	15 5¼	16 8⅝	18 0⅛
195	11 7⅝	12 11⅛	14 2⅝	15 6⅛	16 9¾	18 1¼

Table of the Diameter of Wheels at the Pitch Circle—Continued.

Number of Teeth.	Pitch of the Teeth.					
	inches. 2¼.	inches. 2½.	inches. 2¾.	inches. 3.	inches. 3¼.	inches. 3½.
196	11 8⅜	12 11⅞	14 3½	15 7⅛	16 10¾	18 2⅜
197	11 9	13 0¾	14 4⅜	15 8⅛	16 11¾	18 3½
198	11 9⅞	13 1½	14 5⅜	15 9	17 0⅞	18 4½
199	11 10½	13 2⅜	14 6⅛	15 10	17 1⅞	18 5⅝
200	11 11⅛	13 3⅛	14 7	15 10⅞	17 2⅞	18 6¾
201	11 11⅞	13 3⅞	14 7⅞	15 11⅞	17 3⅞	18 7⅞
202	12 0⅝	13 4¾	14 8⅞	16 0⅞	17 4⅞	18 9
203	12 1⅜	13 5⅜	14 9⅝	16 1⅞	17 6	18 10
204	12 2	13 6¼	14 10½	16 2⅞	17 7	18 11⅛
205	12 2⅞	13 7⅛	14 11⅜	16 3¾	17 8	19 0¼
206	12 3½	13 7⅞	15 0⅜	16 4⅝	17 9⅛	19 1⅜
207	12 4¼	13 8⅝	15 1⅛	16 5⅝	17 10⅛	19 2½
208	12 4⅞	13 9⅜	15 2	16 6⅝	17 11⅛	19 3⅝
209	12 5⅝	13 10¼	15 2⅞	16 7½	18 0¼	19 4¾
210	12 6⅜	13 11⅛	15 3⅞	16 8½	18 1¼	19 5⅞
211	12 7⅛	13 11⅞	15 4⅝	16 9⅜	18 2¼	19 7
212	12 7⅞	14 0⅝	15 5½	16 10⅜	18 3⅛	19 8⅛
213	12 8½	14 1⅜	15 6⅜	16 11⅜	18 4⅜	19 9¼
214	12 9¼	14 2⅛	15 7⅜	17 0⅜	18 5⅞	19 10⅜
215	12 9⅞	14 3	15 8⅛	17 1⅜	18 6⅝	19 11½
216	12 10⅝	14 3⅞	15 9	17 2¼	18 7⅝	20 0⅝
217	12 11⅜	14 4⅝	15 9⅞	17 3⅛	18 8⅝	20 1¾
218	13 0	14 5⅜	15 10⅞	17 4⅛	18 9½	20 2⅞
219	13 0⅞	14 6¼	15 11⅝	17 5⅛	18 10½	20 4
220	13 1½	14 7	16 0½	17 6	18 11½	20 5⅛
221	13 2¼	14 7⅞	16 1⅜	17 7	19 0⅝	20 6¼
222	13 2⅞	14 8⅝	16 2⅜	17 7⅞	19 1⅝	20 7⅜
223	13 3⅝	14 9⅜	16 3⅛	17 8⅞	19 2⅞	20 8½
224	13 4⅜	14 10¼	16 4	17 9⅞	19 3¾	20 9⅝
225	13 5⅛	14 11	16 4⅞	17 10⅞	19 4¾	20 10¾
226	13 5⅞	14 11⅞	16 5⅞	17 11⅞	19 5¾	20 11⅞
227	13 6½	15 0⅝	16 6⅝	18 0¾	19 6⅞	21 0⅞
228	13 7¼	15 1⅜	16 7½	18 1⅝	19 7⅞	21 2
229	13 8	15 2⅛	16 8⅜	18 2⅝	19 8⅞	21 3⅛
230	13 8⅝	15 3	16 9⅜	18 3⅝	19 9⅞	21 4¼
231	13 9⅜	15 3⅞	16 10⅛	18 4½	19 10⅞	21 5⅜
232	13 10⅛	15 4⅝	16 11	18 5½	20 0	21 6⅜
233	13 10⅞	15 5⅜	16 11⅞	18 6⅜	20 1	21 7½
234	13 11½	15 6⅛	17 0⅞	18 7⅜	20 2	21 8⅝
235	14 0¼	15 7	17 1⅝	18 8⅜	20 3	21 9⅞
236	14 1	15 7⅞	17 2½	18 9⅜	20 4⅛	21 10⅞
237	14 1⅝	15 8½	17 3⅜	18 10⅜	20 5⅛	22 0
238	14 2⅜	15 9⅜	17 4⅜	18 11¼	20 6¼	22 1⅛
239	14 3⅛	15 10⅛	17 5⅛	19 0⅛	20 7¼	22 2¼
240	14 3⅞	15 10⅞	17 6	19 1⅛	20 8¼	22 3⅝

Table of the Diameter of Wheels at the Pitch Circle—Continued.

Number of Teeth.	Pitch of the Teeth. inches. 2¼	inches. 2½.	inches. 2¾.	inches. 3.	inches. 3¼.	inches. 3½.
241	14 4½	15 11⅞	17 6⅞	19 2⅞	20 9⅜	22 4⅜
242	14 5¼	16 0½	17 7⅝	19 3⅞	20 10⅜	22 5⅝
243	14 6	16 1⅜	17 8⅝	19 4⅝	20 11⅜	22 6⅝
244	14 6¾	16 2⅛	17 9½	19 5½	21 0⅜	22 7⅞
245	14 7⅜	16 2⅞	17 10¾	19 6⅜	21 1⅝	22 8⅞
246	14 8⅛	16 3¾	17 11⅜	19 7⅜	21 2⅝	22 10
247	14 8⅞	16 4½	18 0⅛	19 8⅛	21 3½	22 11⅛
248	14 9⅝	16 5⅛	18 1	19 9	21 4½	23 0¼
249	14 10⅜	16 6⅛	18 1⅞	19 9⅞	21 5½	23 1⅛
250	14 11	16 6⅞	18 2⅞	19 10⅞	21 6⅝	23 2⅛
251	14 11¾	16 7⅝	18 3⅝	19 11⅝	21 7⅝	23 3⅝
252	15 0⅜	16 8½	18 4½	20 0⅝	21 8⅝	23 4¾
253	15 1⅛	16 9¼	18 5⅜	20 1½	21 9⅝	23 5⅞
254	15 1⅞	16 10⅛	18 6⅜	20 2½	21 10¾	23 6⅞
255	15 2⅝	16 10⅞	18 7⅛	20 3½	21 11¾	23 8
256	15 3⅜	16 11⅞	18 8	20 4⅜	22 0⅞	23 9¼
257	15 4	17 0½	18 8⅞	20 5⅜	22 1⅞	23 10⅜
258	15 4¾	17 1⅜	18 9⅞	20 6⅜	22 2⅞	23 11⅜
259	15 5⅜	17 2⅛	18 10¼	20 7 3/3	22 3⅞	24 0½
260	15 6¼	17 2⅞	18 11½	20 8¼	22 4⅞	24 1⅝
261	15 6⅞	17 3⅝	19 0⅜	20 9⅛	22 6	24 2¾
262	15 7⅝	17 4⅜	19 1⅜	20 10⅛	22 7	24 3⅞
263	15 8⅜	17 5¼	19 2⅛	20 11⅛	22 8	24 5
264	15 9	17 6	19 3	21 0⅛	22 9⅛	24 6⅛
265	15 9¾	17 6⅞	19 3⅞	21 1	22 10⅛	24 7¼
266	15 10⅝	17 7⅝	19 4⅞	21 2	22 11⅛	24 8⅜
267	15 11⅛	17 8⅜	19 5⅝	21 2⅞	23 0¼	24 9⅜
268	15 11⅞	17 9¼	19 6½	21 3⅞	23 1¼	24 10½
269	16 0⅝	17 10	19 7⅜	21 4⅞	23 2¼	24 11⅝
270	16 1⅜	17 10⅞	19 8⅜	21 5⅞	23 3⅜	25 0⅞
271	16 2	17 11⅝	19 9⅜	21 6¾	23 4⅜	25 1⅞
272	16 2¾	18 0⅜	19 10	21 7⅝	23 5⅜	25 3
273	16 3½	18 1⅛	19 10⅞	21 8⅝	23 6⅜	25 4⅛
274	16 4⅛	18 2	19 11⅞	21 9⅝	23 7⅜	25 5¼
275	16 4⅞	18 2⅞	20 0⅝	21 10⅝	23 8⅜	25 6⅜
276	16 5⅜	18 3⅝	20 1½	21 11½	23 9½	25 7⅜
277	16 6⅜	18 4⅜	20 2⅜	22 0½	23 10½	25 8⅝
278	16 7	18 5¼	20 3⅜	22 1⅜	25 11½	25 9⅝
279	16 7⅞	18 6	20 4⅛	22 2⅜	24 0⅝	25 10⅞
280	16 8½	18 6⅞	20 5	22 3⅜	24 1⅝	25 11⅞
281	16 9¼	18 7⅝	20 5⅞	22 4⅜	24 2⅝	26 1
282	16 9⅞	18 8⅜	20 6⅞	22 5¼	24 3¾	26 2⅛
283	16 10⅝	18 9⅛	20 7⅝	22 6⅛	24 4¾	26 3¼
284	16 11⅜	18 9⅞	20 8½	22 7⅛	24 5¾	26 4⅜
285	17 0⅛	18 10¾	20 9⅜	22 8⅛	24 6⅞	26 5½

Table of the Diameter of Wheels at the Pitch Circle—Continued.

Number of Teeth.	Pitch of the Teeth.					
	inches. 2¼.	inches. 2½.	inches. 2¾.	inches. 3.	inches. 3¼.	inches. 3½.
286	17 0⅞	18 11½	20 10⅜	22 9	24 7⅞	26 6⅝
287	17 1½	19 0⅜	20 11⅝	22 10	24 8⅞	26 7⅝
288	17 2¼	19 1⅛	21 0	22 11	24 9⅞	26 8⅞
289	17 2⅞	19 1⅞	21 0⅞	22 11⅞	24 10⅞	26 9⅞
290	17 3⅝	19 2⅜	21 1⅞	23 0⅞	25 0	26 11
291	17 4⅜	19 3½	21 2¾	23 1⅞	25 1	27 0⅛
292	17 5⅛	19 4½	21 3⅝	23 2⅞	25 2	27 1⅜
293	17 5⅞	19 5⅛	21 4⅜	23 3¾	25 3⅛	27 2⅜
294	17 6½	19 5⅞	21 5¼	23 4⅝	25 4⅜	27 3½
295	17 7¼	19 6⅜	21 6⅛	23 5⅝	25 5⅛	27 4⅝
296	17 7⅞	19 7½	21 7	23 6⅝	25 6¼	27 5¾
297	17 8⅝	19 8⅜	21 7⅞	23 7⅝	25 7¼	27 6⅞
298	17 9⅜	19 9⅛	21 8⅞	23 8½	25 8¼	27 7⅞
299	17 10⅛	19 9⅞	21 9⅝	23 9½	25 9⅜	27 9⅛
300	17 10⅞	19 10⅝	21 10⅝	23 10⅝	25 10⅜	27 10¼

Weight of Cast Iron Balls from 1 to 12 Inches Diameter.

Size.	Wt.	Size.	Wt.	Size.	Wt.	Size.	Wt.	Size.	Wt.
Inch.	lbs.	Inch.	lbs.	Inch.	lbs.	Inch.	lbs.	Inch.	lbs.
1	.136	3½	5.84	6	29.45	8½	83.73	11	181.48
1½	.460	4	8.72	6½	37.44	9	99.4	11½	207.37
2	1.09	4½	12.42	7	46.76	9½	116.9	12	235.62
2½	2.13	5	17.04	7½	57.52	10	136.35		
3	3.68	5½	22.68	8	69.81	10½	157.84		

Weight of Cast Iron Pipes 12 Inches Long, from ¼ to 1¼ Inch Thick.

Diam. of Bore.	Inch. ¼	Inch ⅜	Inch. ½	Inch. ⅝	Inch. ¾	Inch. ⅞	Inch. 1	Inch. 1⅛	Inch. 1¼
Inch.	lbs.	lbs.	lbs.	lbs.	lbs.	lbs.	lbs.	lbs.	lbs.
1	3.06	5.06	7.36	9.97	12.89	16.11	19.63		
1¼	3.68	5.98	8.59	11.51	14.73	18.25	22.09		
1½	4.29	6.9	9.82	13.04	16.56	20.4	24.54	28.99	33.74
1¾	4.91	7.83	11.05	14.57	18.41	22.55	27.	31.75	36.76
2	5.53	8.75	12.27	16.11	20.25	24.7	29.45	34.46	39.89
2¼	6.14	9.66	13.5	17.64	22.09	26.84	31.85	37.28	42.95
2½	6.74	10.58	14.72	19.17	23.92	28.93	34.36	40.03	46.02
2¾	7.36	11.5	15.95	20.7	25.71	31.14	36.81	42.8	49.08
3	7.98	12.43	17.18	22.19	27.62	33.29	39.28	45.56	52.16
3¼	8.59	13.34	18.35	23.78	29.45	35.44	41.72	48.32	55.22
3½	9.2	14.21	19.64	25.31	31.3	37.58	44.18	51.08	58.29
3¾	9.76	15.19	20.86	26.85	33.13	39.73	46.63	53.84	61.36
4	10.44	16.11	22.1	28.38	34.98	41.88	49.09	56.61	64.43
4¼	11.1	17.08	23.37	29.97	36.87	44.08	51.6	59.42	67.55
4½	11.66	17.94	24.54	31.44	38.65	46.17	53.99	62.12	70.56
4¾	12.27	18.87	25.77	32.98	40.5	48.32	56.45	64.89	73.63
5	12.88	19.78	26.99	34.51	42.33	50.46	58.9	67.64	76.69
5¼	13.5	20.71	28.23	36.05	44.18	52.62	61.36	70.41	79.77
5½	14.11	21.63	29.45	37.58	46.02	54.76	63.81	73.17	82.84
5¾	14.73	22.55	30.68	39.12	47.86	56.91	66.27	75.94	85.91

Weight of Cast Iron Pipes 12 *Inches Long, from* ¼ *to* 1¼ *Inch Thick*—Cont.

Diam. of Bore.	Inch, ¼	Inch, ⅜	Inch, ½	Inch, ⅝	Inch, ¾	Inch, ⅞	Inch, 1	Inch, 1⅛	Inch, 1¼
Inch.	lbs.	lbs.	lbs.	lbs.	lbs.	lbs.	lbs.	lbs.	lbs.
6	15.34	23.47	31.91	40.65	49.7	59.06	68 73	78.7	88.75
6¼	15.95	24.39	33.13	42.18	51.54	61.21	71.18	81.23	92.04
6½	16.57	25.31	34.36	43.72	53.39	63.36	73.41	84.22	95.1
6¾	17.18	26.23	35.59	45.26	55.23	65.28	76.09	86.97	98.18
7	17.79	27.15	36.82	46.79	56.84	67.65	78.53	89.74	101.24
7¼	18.41	28.08	38.05	48.1	58.91	69.79	81.	92.5	104.31
7½	19.03	29.	39.05	49.86	60.74	71.95	83.45	95.26	107.38
7¾	19.64	29.69	40.5	51.38	62.59	74.09	85.9	98.02	110.45
8	20.02	30.83	41.71	52.92	64.42	76.23	88.35	100.78	113.51
8¼	20.86	31.74	42.95	54.45	66.26	78.38	90.81	103.54	116.58
8½	21.69	32.9	44.4	56.21	68.33	80.76	93.49	106.53	119.87
8¾	22.09	33.59	45.4	57.52	69.95	82.68	95.72	109.06	122.72
9	22.71	34.52	46.64	59.07	71.8	84.84	98.18	111.84	125.8
9¼	23.31	35.43	47.86	60.59	73.63	86.97	100.63	114.59	128.85
9½	23.93	36.36	49.09	62.13	75.47	89.13	103.09	117.35	131.93
9¾	24.55	37.28	50.32	63.66	77.32	91.28	105.54	120.12	134.99
10	25.16	38.2	51.54	65.2	79.16	93.42	108.	122.87	138.06
10¼	25.77	39.11	52.77	66.73	80.99	95.57	110.44	125.63	141.12
10½	26.38	40.04	54.	68.26	82.84	97.71	112.9	128.39	144.19
10¾	27.	40.96	55.22	69.8	84.67	99.86	115.35	131.15	147.26
11	27.62	41.88	56.46	71.33	86.52	102.01	117.81	133.92	150.33
11¼	28.22	42.8	57.67	72.86	88.35	104.15	120.26	136.67	153.4
11½	28.84	43.71	58.9	74.39	90.19	106.3	122.71	139.44	156.44
11¾	29.45	44.64	60.13	75.93	92.04	108.45	125.18	142.18	159.54
12	30.06	45.55	61.35	77.46	93.6	110.6	127.6	144.96	162.6

Weight of Cast Iron Pipes 12 *Inches Long, from* 1⅜ *to* 1½ *Inch Thick.*

D. of B.	1⅜ Inch.	1½ Inch.	D. of B.	1⅜ Inch.	1½ Inch.	D. of B.	1⅜ Inch.	1½ Inch.
Inch.	lbs,	lbs.	Inch.	lbs.	lbs.	Inch.	lbs.	lbs.
2¼	48.94	55.22	5¾	95.96	106.77	9	140.06	154.64
2½	52.30	58.9	6	99.56	110.44	9¼	143.43	158.3
2¾	55.68	62.58	6¼	102.92	114.13	9½	146.8	161.99
3	59.06	66.27	6½	106.31	117.81	9¾	150.18	165.67
3¼	62.43	69.95	6¾	109.68	121.49	10	153.55	169.35
3½	65.81	73.63	7	113.05	125.17	10¼	156.92	173.03
3¾	69.18	77.31	7¼	116.43	128.86	10½	160.3	176.71
4	72.56	81.	7½	119.81	132.54	10¾	163.67	180.4
4¼	75.99	84.73	7¾	123.18	136.22	11	167.06	184.06
4½	79.3	88.35	8	126.55	139.89	11¼	170.4	187.76
4¾	82.68	92.04	8¼	129.92	143.58	11½	173.8	191.44
5	86.05	95.72	8½	133.53	147.49	11¾	177.18	195.12
5¼	89.44	99.41	8¾	136.68	150.94	12	180.54	198.8
5½	92.81	102.86						

Round Cast Iron Twelve Inches Long.

Size.	Weight.	Size.	Weight.	Size.	Weight.	Size.	Weight.	Size.	Weight.
Inch.	lbs.	Inch.	lbs.	Inch.	lbs.	Inch.	lbs.	Inch.	lbs.
½	.61	2¼	12.42	4	39.27	5¾	81.14	9	198.79
⅝	.95	2⅜	13.84	4⅛	41.76	5⅞	84.71	9¼	210.
¾	1.38	2½	15.33	4¼	44.27	6	88.35	9½	221.5
⅞	1.87	2⅝	16.91	4⅜	46.97	6¼	95.87	9¾	233.34
1	2.45	2¾	18.56	4½	49.7	6½	103.69	10	245.43
1⅛	3.1	2⅞	20.28	4⅝	52.5	6¾	111.82	10¼	257.86
1¼	3.83	3	22.08	4¾	55.37	7	120.26	10½	270.59
1⅜	4.64	3⅛	23.96	4⅞	58.32	7¼	129.	10¾	283.63
1½	5.52	3¼	25.92	5	61.35	7½	138.05	11	296.97
1⅝	6.48	3⅜	27.95	5⅛	64.46	7¾	147.41	11¼	310.63
1¾	7.51	3½	30.06	5¼	67.64	8	157.08	11½	324.59
1⅞	8.62	3⅝	32.25	5⅜	70.09	8¼	167.05	11¾	338.85
2	9.81	3¾	34.51	5½	74.24	8½	177.1	12	353.43
2⅛	11.08	3⅞	36.85	5⅝	77.65	8¾	187.91		

Square Cast Iron Twelve Inches Long.

Size.	Weight.	Size.	Weight.	Size.	Weight.	Size.	Weight.	Size.	Weight.
Inch.	lbs.	Inch.	lbs.	Inch.	lbs.	Inch.	lbs.	Inch.	lbs.
½	.78	2¼	15.81	4	50.	5¾	103.32	9	253.12
⅝	1.22	2⅜	17.62	4⅛	53.14	5⅞	107.86	9¼	267.38
¾	1.75	2½	19.53	4¼	56.44	6	112.5	9½	282.
⅞	2.39	2⅝	21.53	4⅜	59.81	6¼	122.08	9¾	297.07
1	3.12	2¾	23.63	4½	63.28	6½	132.03	10	312.5
1⅛	3.95	2⅞	25.83	4⅝	66.84	6¾	142.38	10¼	328.32
1¼	4.88	3	28.12	4¾	70.5	7	153.12	10½	344.53
1⅜	5.9	3⅛	30.51	4⅞	74.26	7¼	164.25	10¾	361.13
1½	7.03	3¼	33.	5	78.12	7½	175.78	11	378.12
1⅝	8.25	3⅜	35.59	5⅛	82.08	7¾	187.68	11¼	395.5
1¾	9.57	3½	38.28	5¼	86.13	8	200.	11½	413.28
1⅞	10.98	3⅝	41.06	5⅜	90.28	8¼	212.56	11¾	431.44
2	12.5	3¾	43.94	5½	94.53	8½	225.78	12	450.
2⅛	14.11	3⅞	46.92	5⅝	98.87	8¾	239.25		

Flat Cast Iron Twelve Inches Long, ¼ to 1 Inch Thick.

Width of Iron.	Inch. ¼	Inch. ⅜	Inch. ½	Inch. ⅝	Inch. ¾	Inch. ⅞	Inch. 1
Inch.	lbs.	lbs.	lbs.	lbs.	lbs.	lbs.	lbs.
2	1.56	2.34	3.12	3.9	4.68	5.46	6.25
2¼	1.75	2.63	3.51	4.39	5.27	6.15	7.03
2½	1.95	2.92	3.9	4.88	5.85	6.83	7.81
2¾	2.14	3.22	4.29	5.37	6.44	7.51	8.59
3	2.34	3.51	4.68	5.85	7.03	8.2	9.37
3¼	2.53	3.8	5.07	6.34	7.61	8.88	10.15
3½	2.73	4.1	5.46	6.83	8.2	9.57	10.93
3¾	2.93	4.39	5.85	7.32	8.78	10.25	11.71
4	3.12	4.68	6.25	7.81	9.37	10.93	12.5
4¼	3.32	4.97	6.64	8.3	9.96	11.62	13.28
4½	3.51	5.27	7.03	8.78	10.54	12.3	14.06
4¾	3.71	5.56	7.42	9.27	11.13	12.98	14.84
5	3.9	5.86	7.81	9.76	11.71	13.67	15.62
5¼	4.1	6.15	8.2	10.25	12.3	14.35	16.4
5½	4.29	6.44	8.59	10.74	12.89	15.03	17.18
5¾	4.49	6.73	8.98	11.23	13.46	15.72	17.96
6	4.68	7.03	9.37	11.71	14.06	16.4	18.75

Weight of a Superficial Foot of Cast Iron from ¼ to 2 Inches Thick.

Thickness.	¼	⅜	½	⅝	¾	⅞	1	1⅛
Wt.	lbs. 9.37	lbs. 14.06	lbs. 18.75	lbs. 23.43	lbs. 28.12	lbs. 32.81	lbs. 37.5	lbs. 42.18

Thickness.	1¼	1⅜	1½	1⅝	1¾	1⅞	2	
Wt.	lbs. 46.87	lbs. 51.56	lbs. 56.25	lbs. 60.93	lbs. 65.62	lbs. 70.31	lbs. 75.	

Weight of Square Lead Twelve Inches Long, from 1 to 3 Inches Square.

Size.	1 in.	1⅛	1¼	1⅜	1½	1⅝	1¾	1⅞	2
Wt.	lbs, 4.93	lbs. 6.25	lbs. 7.71	lbs. 9.33	lbs. 11.11	lbs. 13.04	lbs. 15.12	lbs. 17.36	lbs. 19.75

Size.	2⅛	2¼	2⅜	2½	2⅝	2¾	2⅞	3	
Wt.	lbs. 22.29	lbs. 25.	lbs. 27.8	lbs. 30.86	lbs. 34.02	lbs. 37.34	lbs. 40.81	lbs. 44.44	

Weight of Round Lead Twelve Inches Long, from 1 to 3 Inches Diameter.

Size.	1 in.	1⅛	1¼	1⅜	1½	1⅝	1¾	1⅞	2
Wt.	lbs. 3.87	lbs. 4.9	lbs. 6.06	lbs. 7.33	lbs. 8.72	lbs. 10.24	lbs. 11.87	lbs. 13.63	lbs. 15.51

Size.	2⅛	2¼	2⅜	2½	2⅝	2¾	2⅞	3	
Wt.	lbs. 17.51	lbs. 19.63	lbs. 21.8	lbs. 24.24	lbs. 26.72	lbs. 29.33	lbs. 32.05	lbs. 34.9	

Binary and Decimal Fractions.

1/64 =.015625	23/64 =.359375	11/16 =.6875
1/32 =.03125	3/8 =.375	45/64 =.703125
3/64 =.046875	25/64 =.390625	23/32 =.71875
1/16 =.0625	13/32 =.40625	47/64 =.734375
5/64 =.078125	27/64 =.421875	3/4 =.75
3/32 =.09375	7/16 =.4375	49/64 =.765625
7/64 =.109375	29/64 =.453125	25/32 =.78125
1/8 =.125	15/32 =.46875	51/64 =.796875
9/64 =.140625	31/64 =.484375	13/16 =.8125
5/32 =.15625	1/2 =.5	53/64 =.828125
11/64 =.171875	33/64 =.515625	27/32 =.84375
3/16 =.1875	17/32 =.53125	55/64 =.859375
13/64 =.203125	35/64 =.546875	7/8 =.875
7/32 =.21875	9/16 =.5625	57/64 =.890625
15/64 =.234375	37/64 =.578125	29/32 =.90625
1/4 =.25	19/32 =.59375	59/64 =.921875
17/64 =.265625	39/64 =.609375	15/16 =.9375
9/32 =.28125	5/8 =.625	61/64 =.953125
19/64 =.296875	41/64 =.640625	31/32 =.96875
5/16 =.3125	21/32 =.65625	63/64 =.984375
21/64 =.328125	43/64 =.671875	1 =1.000000
11/32 =.34375		

Distances at which to open a 2 ft. Rule to obtain a given Angle.

Angle.	Distance.	Angle.	Distance.	Angle.	Distance.	Angle.	Distance.	Angle.	Distance.
Deg.	Inches	Deg.	Inches	Deg.	Inches	Deg.	Inches	Deg.	Inches
1	.2	19	3.96	37	7.61	55	11.08	73	14.28
2	.42	20	4.17	38	7.81	56	11.27	74	14.44
3	.63	21	4.37	39	8.01	57	11.45	75	14.61
4	.84	22	4.58	40	8.20	58	11.64	76	14.78
5	1.05	23	4.78	41	8.40	59	11.82	77	14.94
6	1.26	24	4.99	42	8.60	60	12.00	78	15.11
7	1.47	25	5.19	43	8.80	61	12.18	79	15.27
8	1.67	26	5.40	44	8.99	62	12.36	80	15.43
9	1.88	27	5.60	45	9.18	63	12.54	81	15.59
10	2.09	28	5.81	46	9.38	64	12.72	82	15.75
11	2.30	29	6.01	47	9.57	65	12.90	83	15.90
12	2.51	30	6.21	48	9.76	66	13.07	84	16.06
13	2.72	31	6.41	49	9.95	67	13.25	85	16.21
14	2.92	32	6.62	50	10.14	68	13.42	86	16.37
15	3.13	33	6.82	51	10.33	69	13.59	87	16.52
16	3.34	34	7.02	52	10.52	70	13.77	88	16.67
17	3.55	35	7.22	53	10.71	71	13.94	89	16.82
18	3.75	36	7.42	54	10.90	72	14.11	90	16.97

French Mètre reduced to Inches.

Mètre.		Inches.
.001587	=	$\frac{1}{16}$
.00317	=	$\frac{1}{8}$
.00476	=	$\frac{3}{16}$
.00635	=	$\frac{1}{4}$
.00794	=	$\frac{5}{16}$
.00952	=	$\frac{3}{8}$
.01111	=	$\frac{7}{16}$
.01270	=	$\frac{1}{2}$
.01429	=	$\frac{9}{16}$
.01587	=	$\frac{5}{8}$
.01746	=	[illegible]
.01905	=	$\frac{3}{4}$
.02064	=	$\frac{13}{16}$
.02222	=	$\frac{7}{8}$
.02381	=	$\frac{15}{16}$
.02540	=	1
.05078	=	2
.0762	=	3
.1016	=	4
.1270	=	5
.1524	=	6
.1778	=	7
.2032	=	8
.2286	=	9
.2540	=	10
.2794	=	11
.3048	=	12

Milli-mètres.		Mètre.		Inches.		Feet.
1	=	.001	=	.03937	=	.00328
2	=	.002	=	.07874	=	.00656
3	=	.003	=	.11811	=	.00984
4	=	.004	=	.15748	=	.01312
5	=	.005	=	.19685	=	.01641
6	=	.006	=	.23622	=	.01969
7	=	.007	=	.2756	=	.02397
8	=	.008	=	.31497	=	.02625
9	=	.009	=	.35434	=	.02953
Centi-mètres.						
1	=	.01	=	.3937	=	.0328
2	=	.02	=	.7874	=	.0656
3	=	.03	=	1.1811	=	.0984
4	=	.04	=	1.5748	=	.1312
5	=	.05	=	1.9685	=	.1641
6	=	.06	=	2.3622	=	.1969
7	=	.07	=	2.756	=	.2397
8	=	.08	=	3.1497	=	.2625
9	=	.09	=	3.5434	=	.2953
Deci-mètres.						
1	=	.1	=	3.9371	=	.3281
2	=	.2	=	7.8742	=	.6562
3	=	.3	=	11.8112	=	.9843
4	=	.4	=	15.7483	=	1.3124
5	=	.5	=	19.6854	=	1.6404
6	=	.6	=	23.6225	=	1.9685
7	=	.7	=	27.5596	=	2.3966
8	=	.8	=	31.4966	=	2.6247
9	=	.9	=	35.4337	=	2.9528

THE MÈTRE = 3.2808992 FEET (ABOUT 39$\frac{3}{8}$ INCHES).

PLATES OF GEAR TEETH

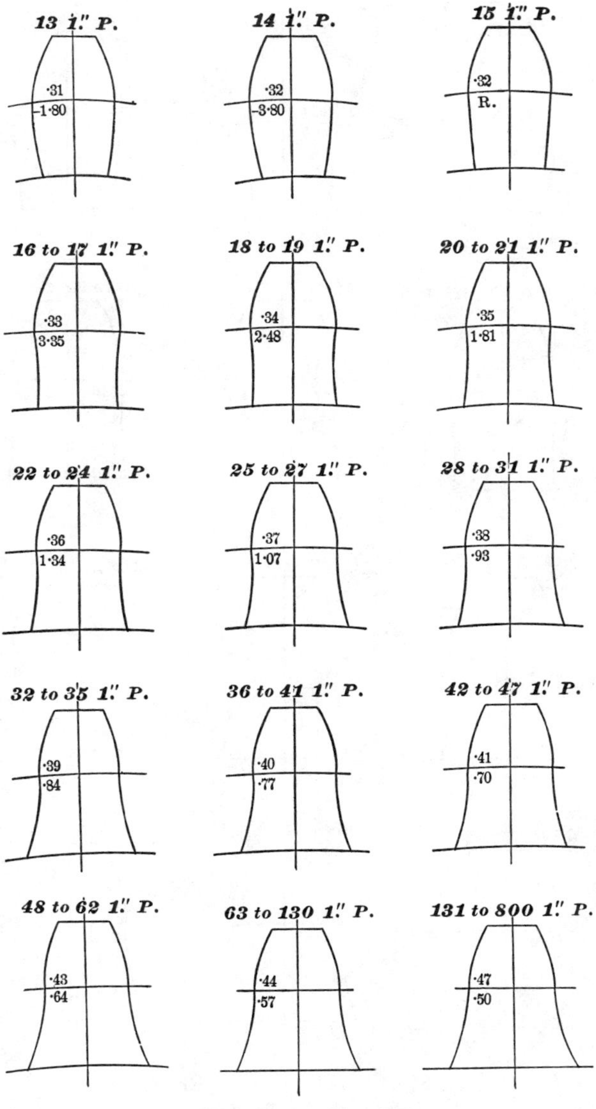

FULL SIZE GEAR TEETH.

From Prof. S. W. Robinson's Templet Odontograph.

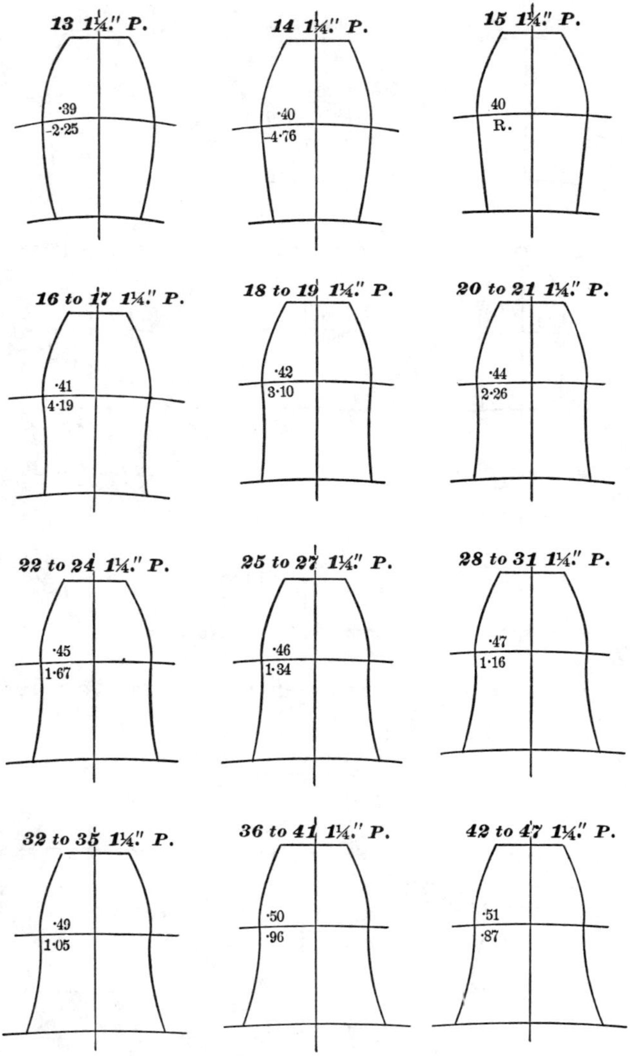

FULL SIZE GEAR TEETH.

From Prof. S. W. Robinson's Templet Odontograph.

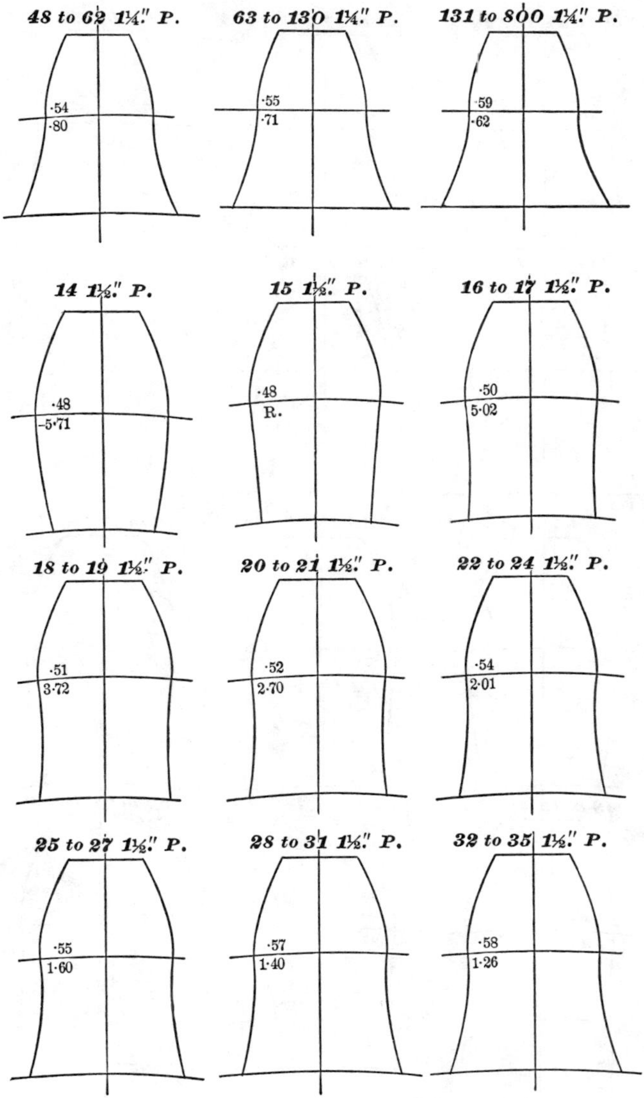

FULL SIZE GEAR TEETH.

From Prof. S. W. Robinson's Templet Odontograph.

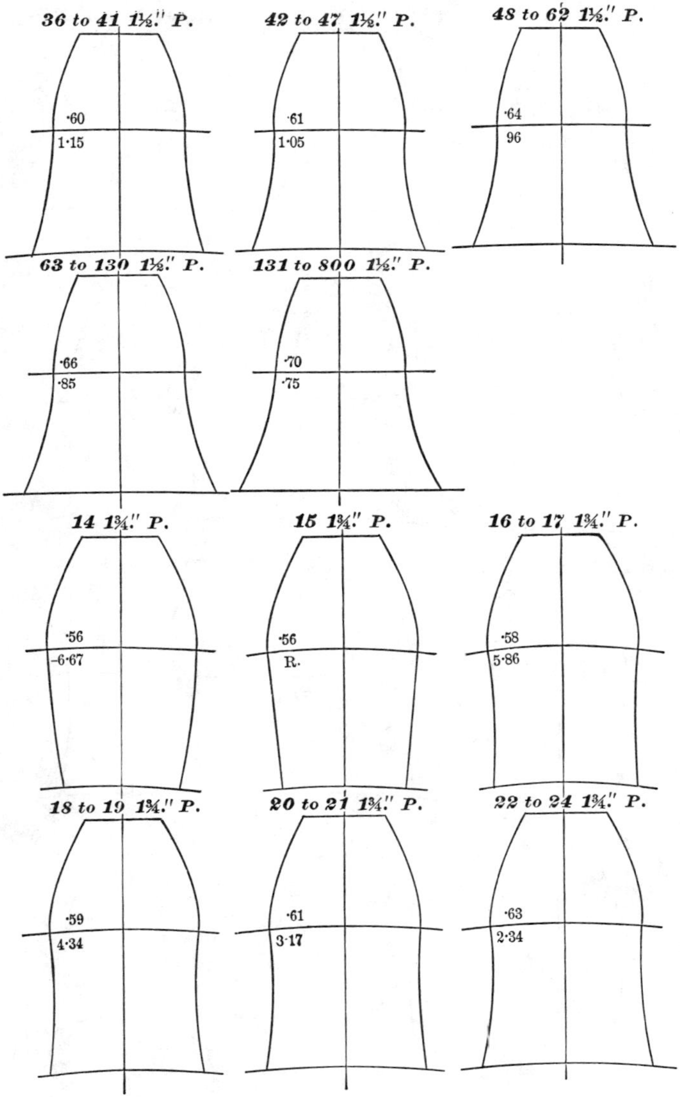

FULL SIZE GEAR TEETH.

From Prof. S. W. Robinson's Templet Odontograph.

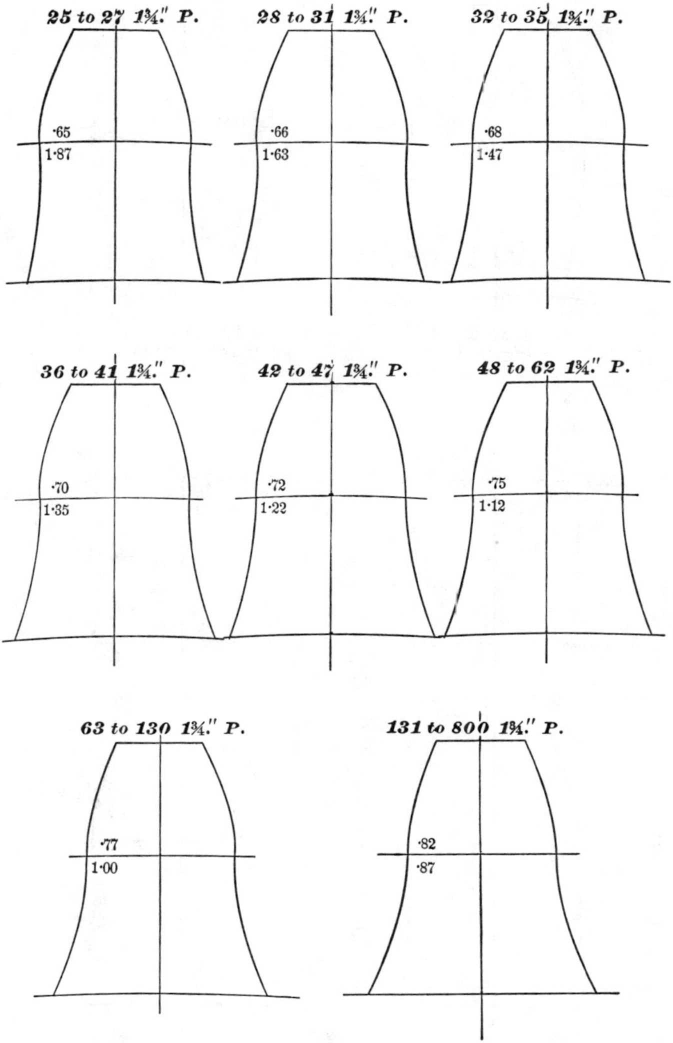

FULL SIZE GEAR TEETH.

From Prof. S. W. Robinson's Templet Odontograph.

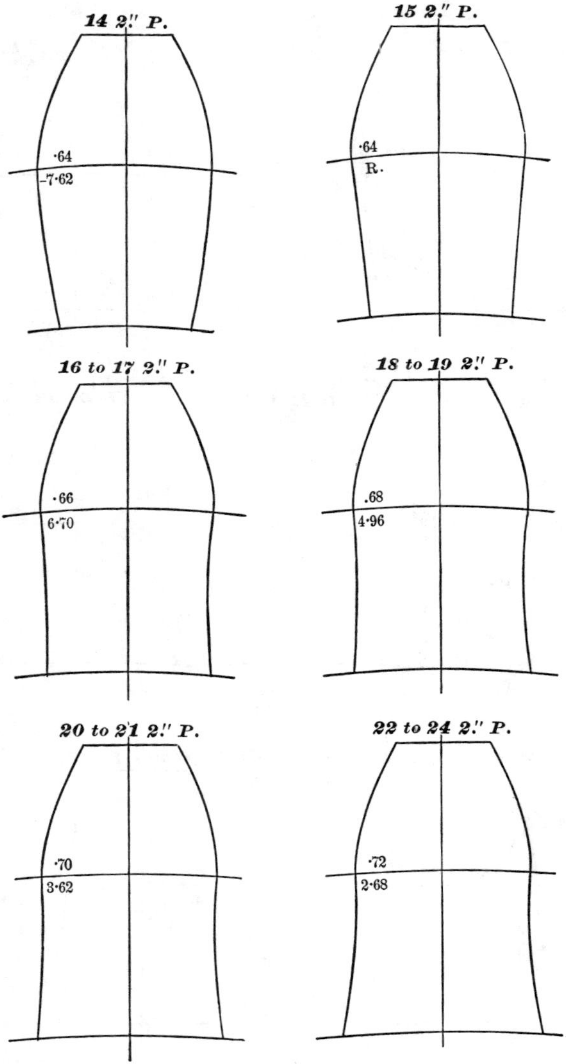

FULL SIZE GEAR TEETH.
From Prof. S. W. Robinson's Templet Odontograph.

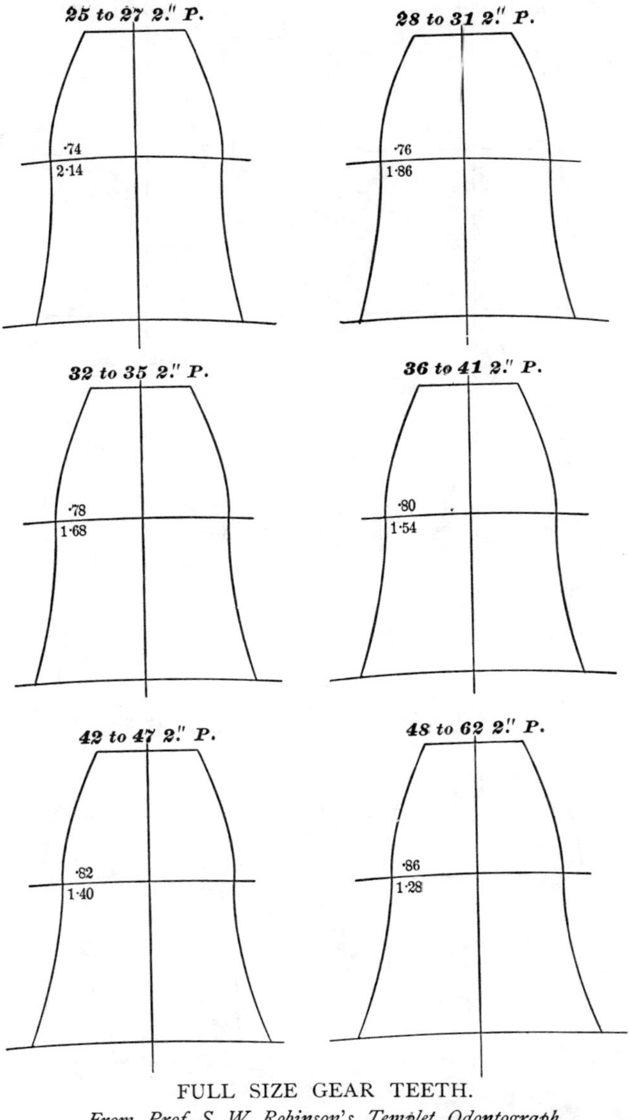

FULL SIZE GEAR TEETH.

From Prof. S. W. Robinson's Templet Odontograph.

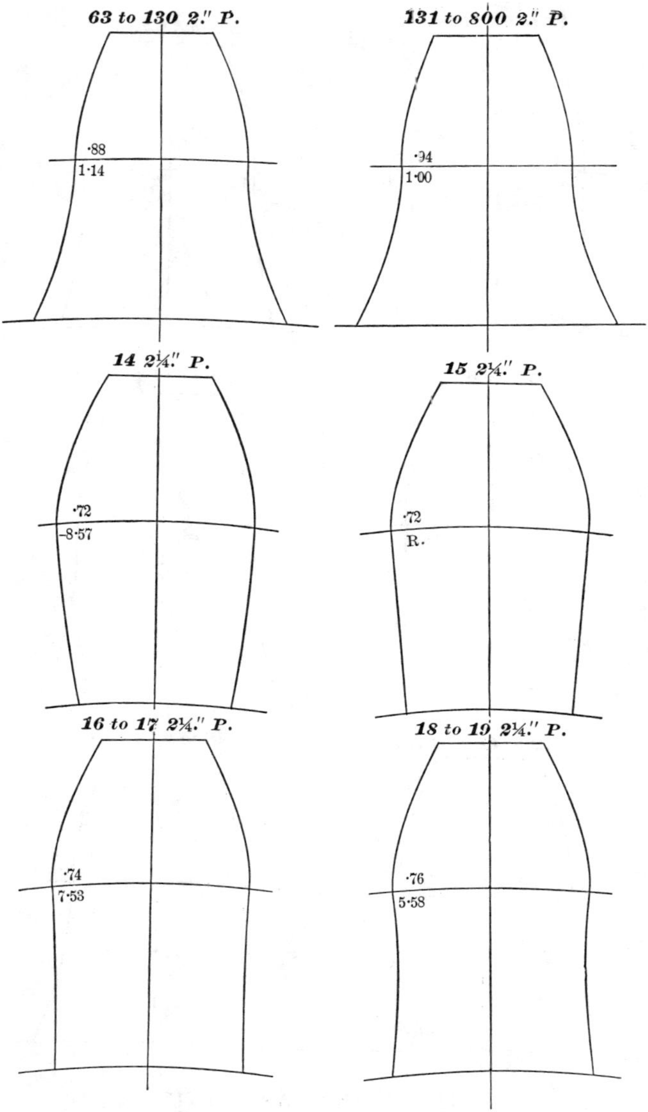

FULL SIZE GEAR TEETH.

From Prof. S. W. Robinson's Templet Odontograph.

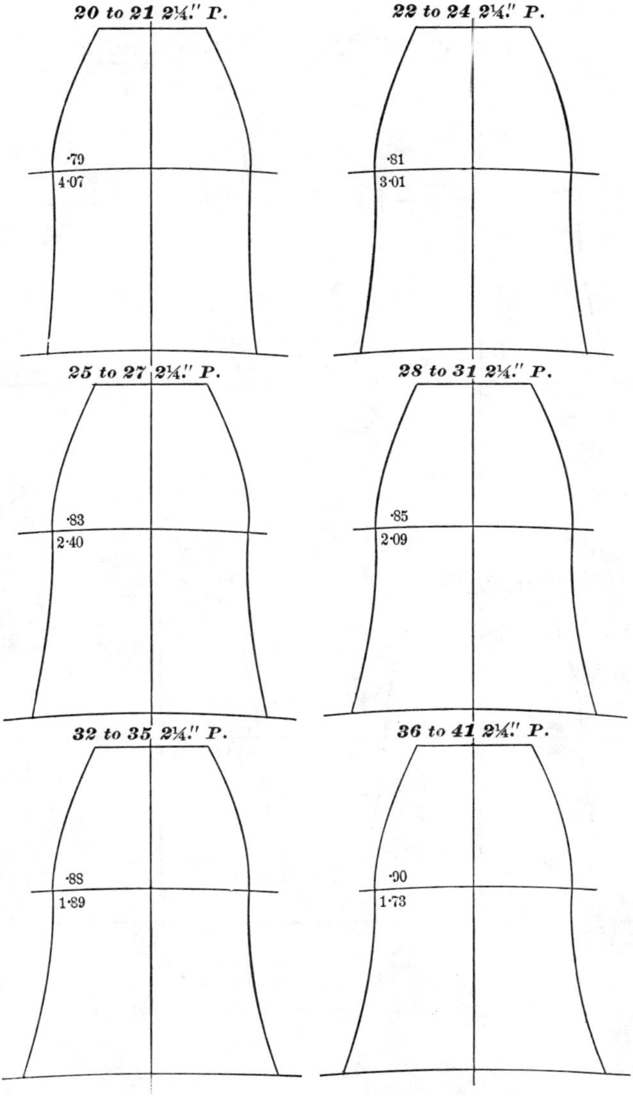

FULL SIZE GEAR TEETH.
From Prof. S. W. Robinson's Templet Odontograph

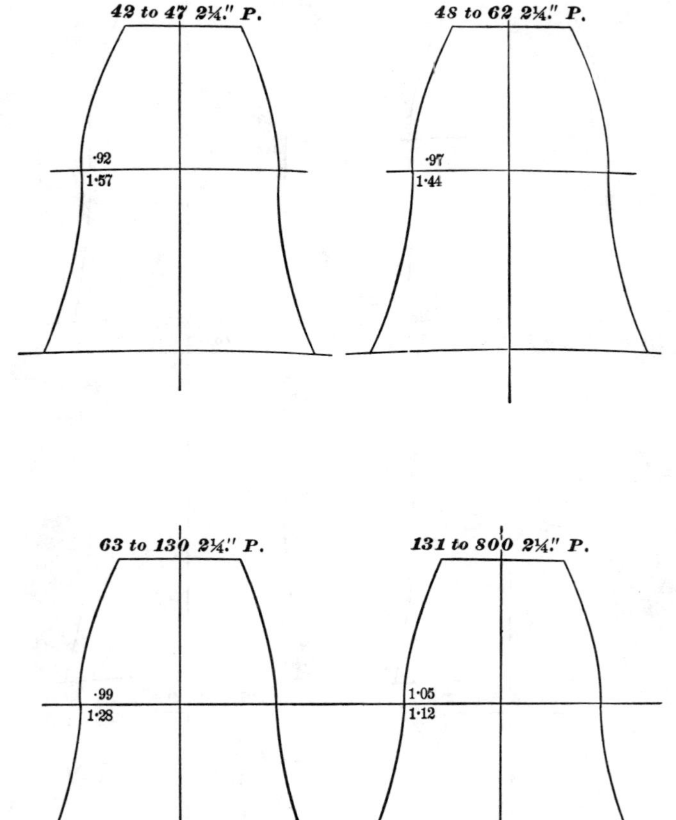

FULL SIZE GEAR TEETH.

From Prof. S. W. Robinson's Templet Odontograph.

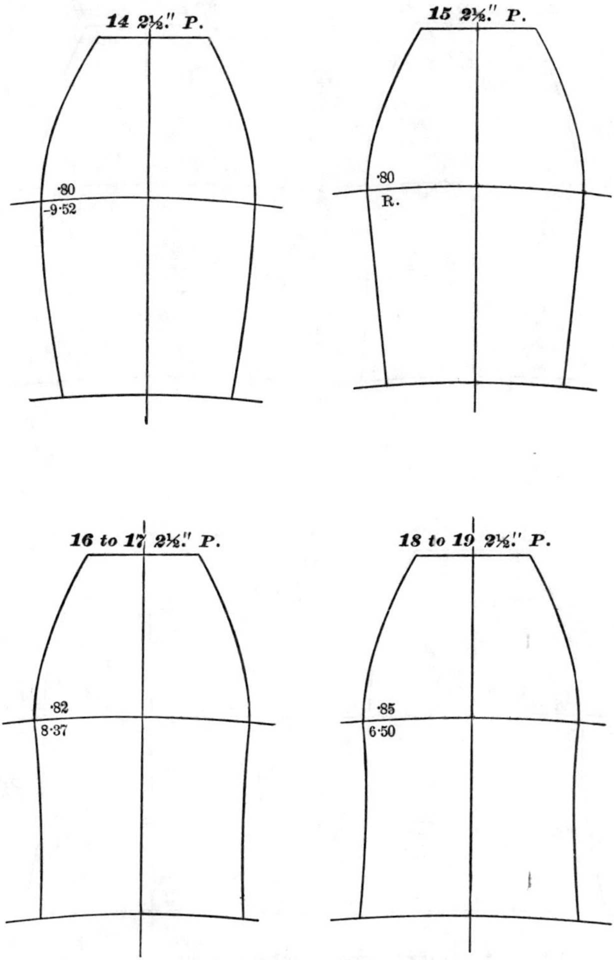

FULL SIZE GEAR TEETH.

From Prof. S. W. Robinson's Templet Odontograph.

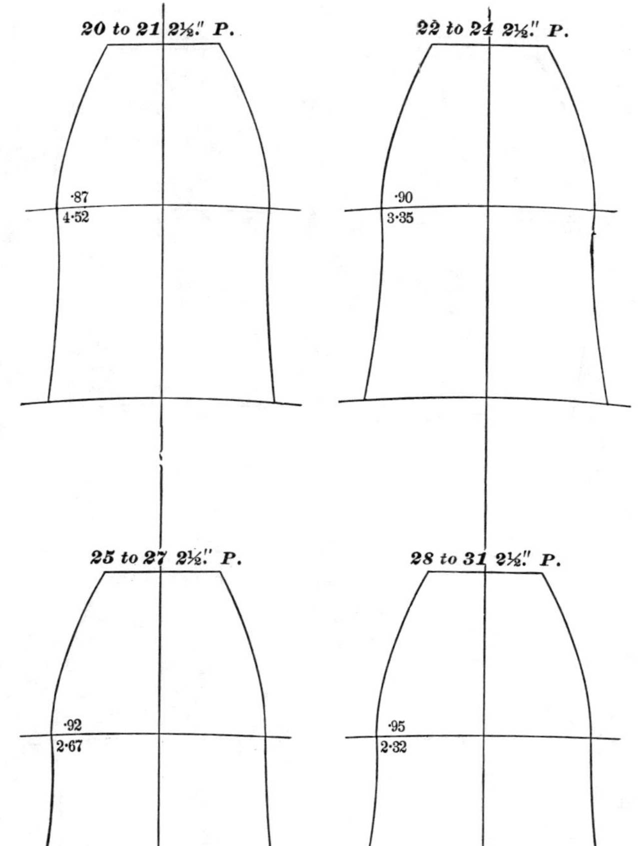

FULL SIZE GEAR TEETH.

From Prof. S. W. Robinson's Templet Odontograph.

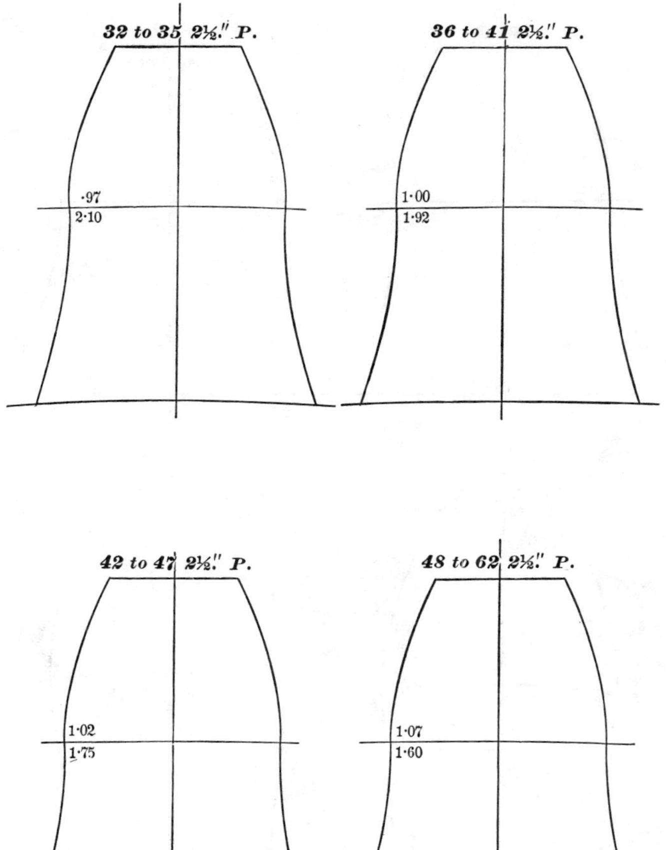

FULL SIZE GEAR TEETH.
From Prof. S. W. Robinson's Templet Odontograph.

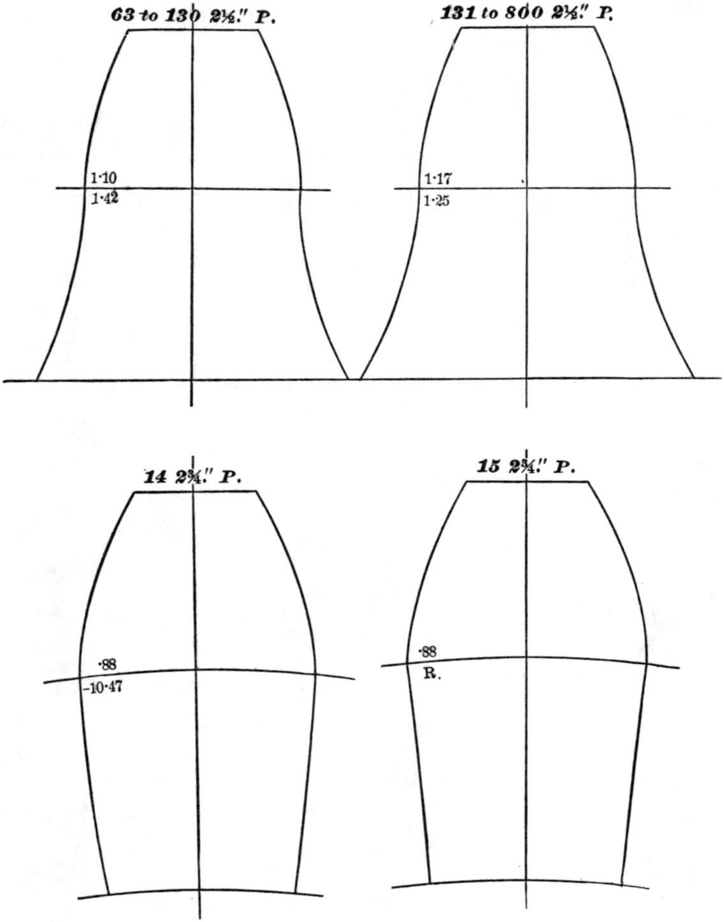

FULL SIZE GEAR TEETH.

From Prof. S. W. Robinson's Templet Odontograph.

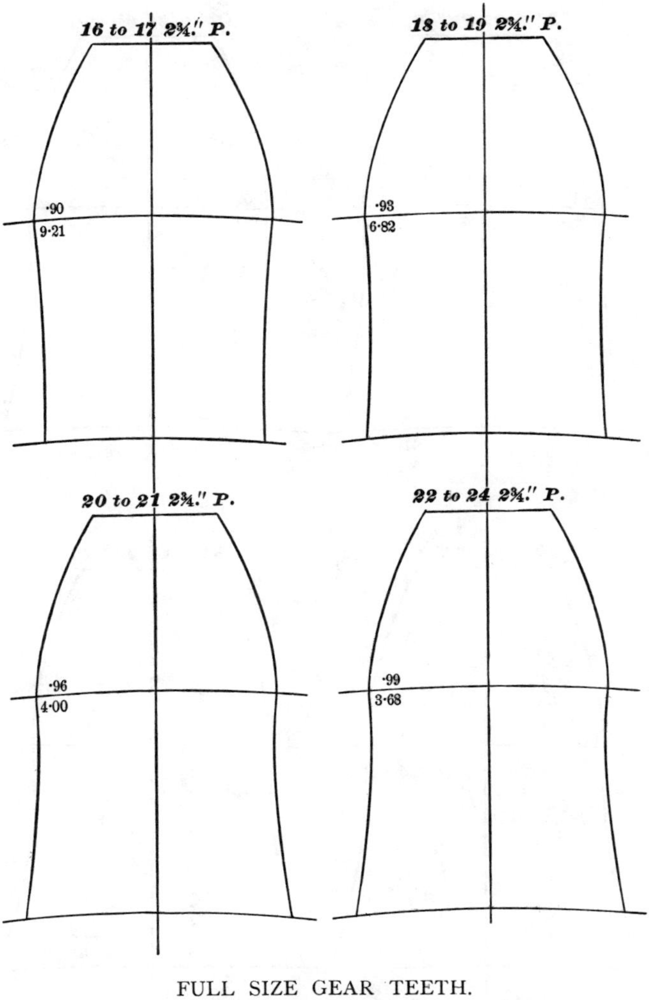

FULL SIZE GEAR TEETH.
From Prof. S. W. Robinson's Templet Odontograph.

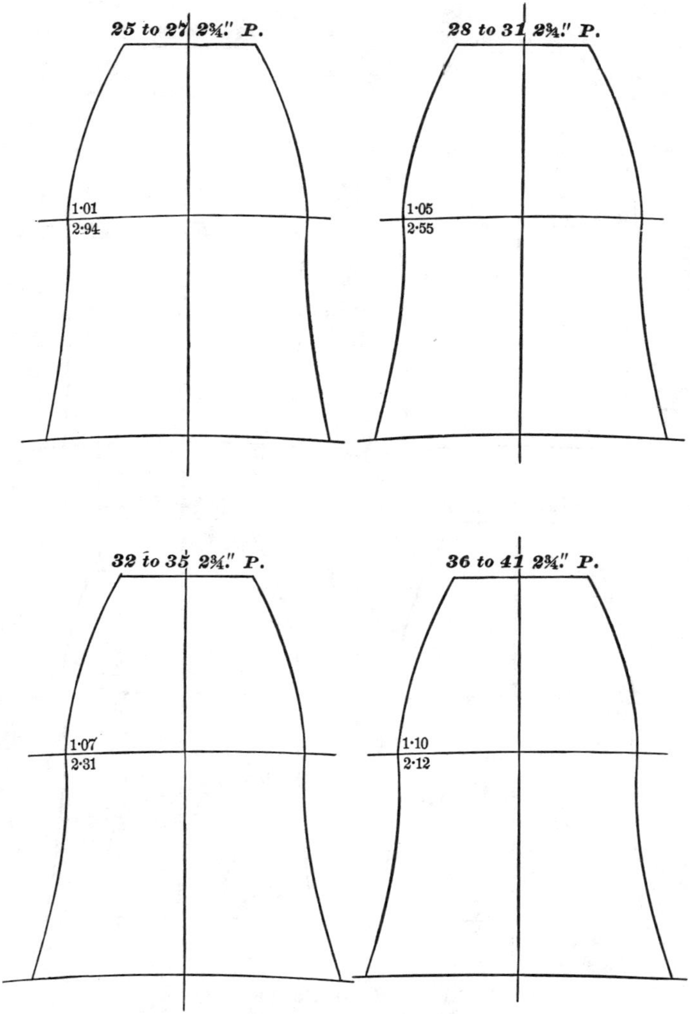

FULL SIZE GEAR TEETH.

From Prof. S. W. Robinson's Templet Odontograph.

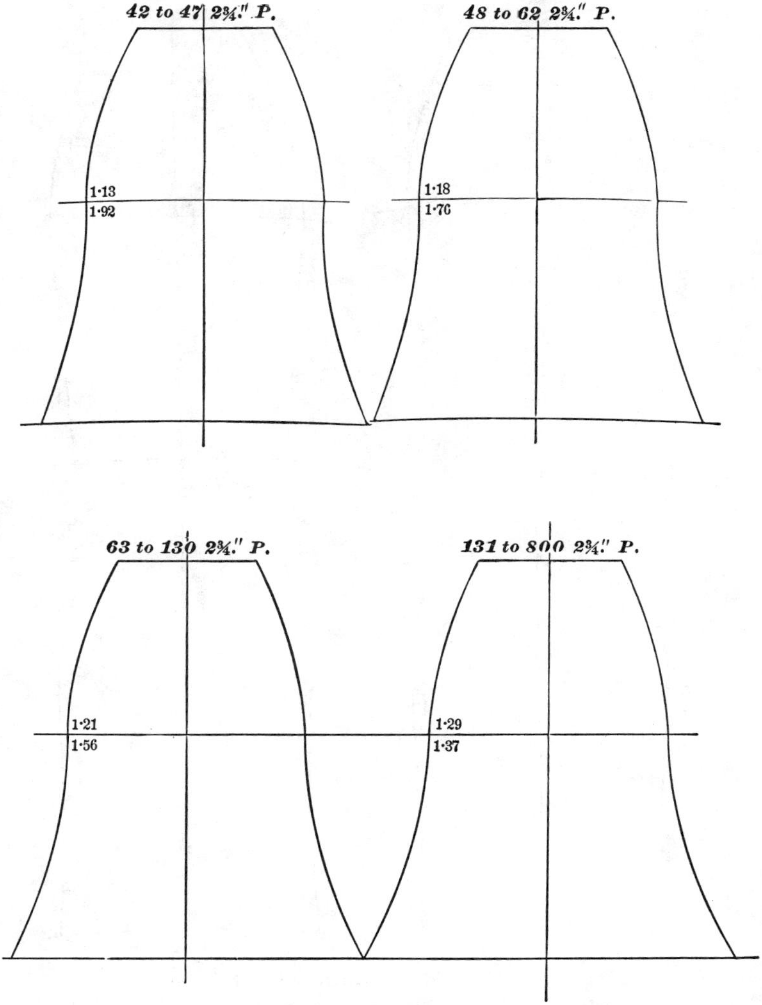

FULL SIZE GEAR TEETH.

From Prof. S. W. Robinson's Templet Odontograph.

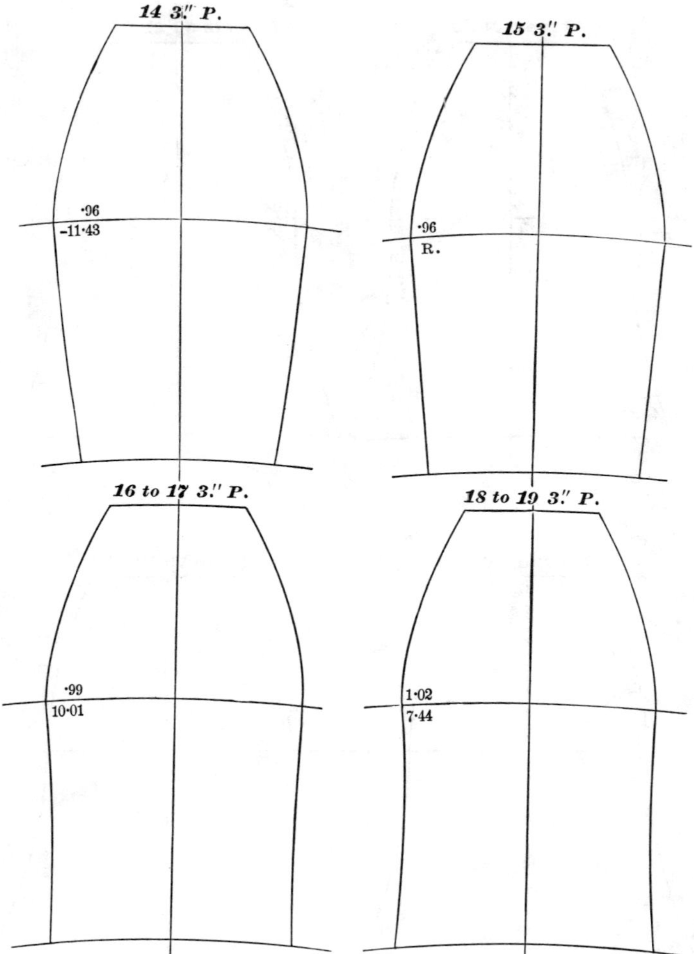

FULL SIZE GEAR TEETH.
From Prof. S. W. Robinson's Templet Odontograph.

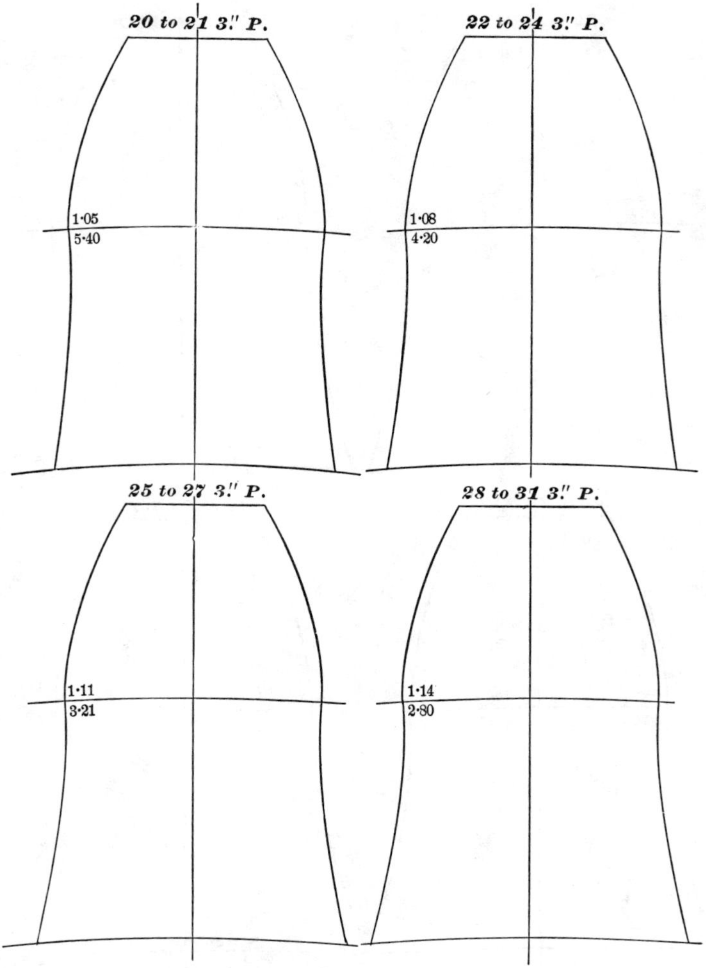

FULL SIZE GEAR TEETH.
From Prof. S. W. Robinson's Templet Odontograph.

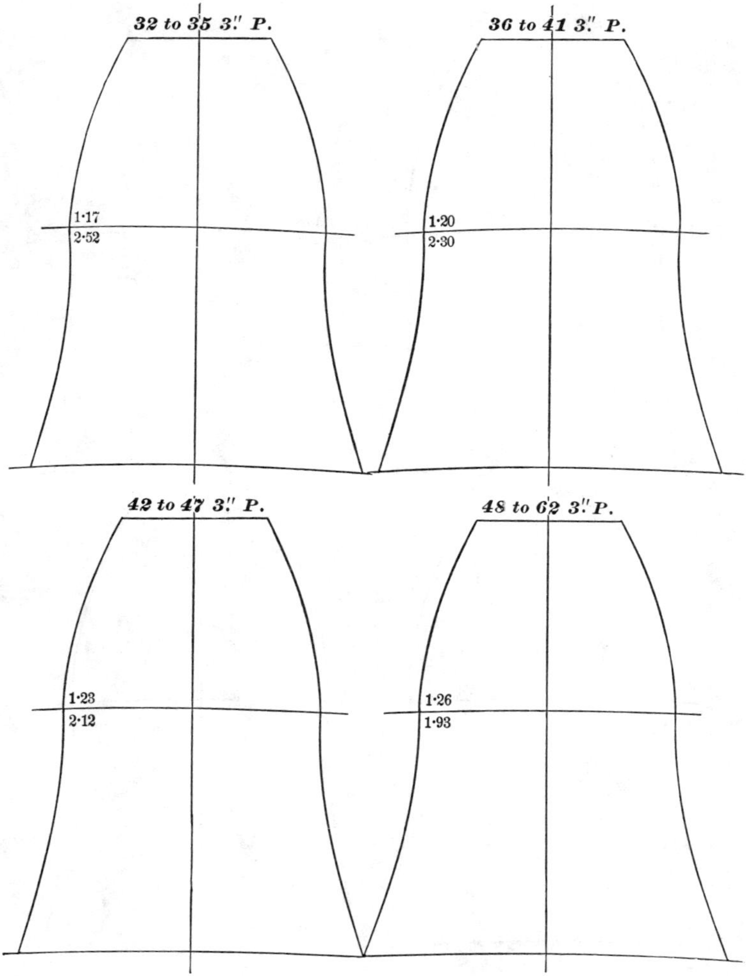

FULL SIZE GEAR TEETH.

From Prof. S. W. Robinson's Templet Odontograph.

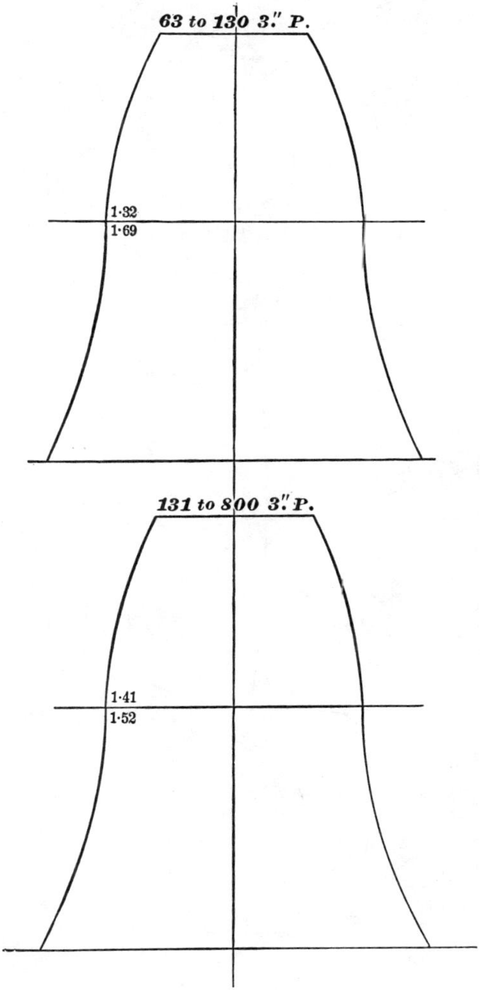

FULL SIZE GEAR TEETH.

From Prof. S. W. Robinson's Templet Odontograph.

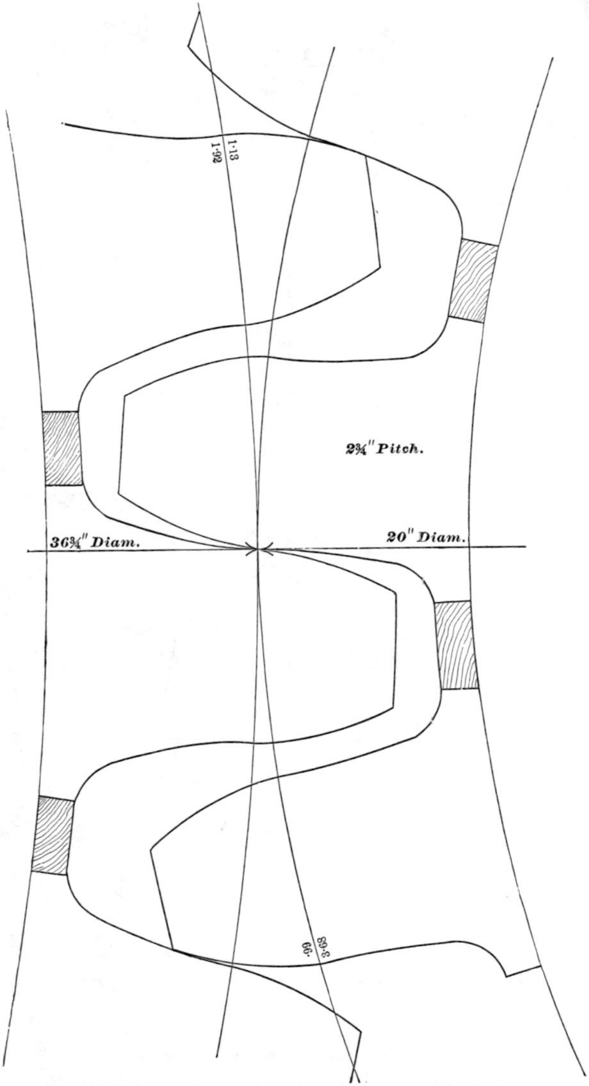

FULL SIZE GEAR TEETH.

From Prof. S. W. Robinson's Templet Odontograph.

SHORT-TITLE CATALOGUE

OF THE

PUBLICATIONS

OF

JOHN WILEY & SONS,

NEW YORK.

LONDON: CHAPMAN & HALL, LIMITED.

ARRANGED UNDER SUBJECTS.

Descriptive circulars sent on application.

Books marked with an asterisk are sold at *net* prices only.

All books are bound in cloth unless otherwise stated.

AGRICULTURE.

CATTLE FEEDING—DAIRY PRACTICE—DISEASES OF ANIMALS—GARDENING, ETC.

Armsby's Manual of Cattle Feeding	12mo,	$1 75
Downing's Fruit and Fruit Trees	8vo,	5 00
Grotenfelt's The Principles of Modern Dairy Practice. (Woll.)	12mo,	2 00
Kemp's Landscape Gardening	12mo,	2 50
Lloyd's Science of Agriculture	8vo,	4 00
Loudon's Gardening for Ladies. (Downing.)	12mo,	1 50
Steel's Treatise on the Diseases of the Dog	8vo,	3 50
" Treatise on the Diseases of the Ox	8vo,	6 00
Stockbridge's Rocks and Soils	8vo,	2 50
Woll's Handbook for Farmers and Dairymen	12mo,	1 50

ARCHITECTURE.

BUILDING—CARPENTRY—STAIRS—VENTILATION, ETC.

Berg's Buildings and Structures of American Railroads	4to,	7 50
Birkmire's American Theatres—Planning and Construction	8vo,	3 00
" Architectural Iron and Steel	8vo,	3 50
Birkmire's Compound Riveted Girders	8vo,	2 00
" Skeleton Construction in Buildings	8vo,	3 00
Carpenter's Heating and Ventilating of Buildings	8vo,	3 00

Downing, Cottages	8vo,	$2 50
" Hints to Architects	8vo,	2 00
Freitag's Architectural Engineering	8vo,	2 50
Gerhard's Sanitary House Inspection	16mo,	1 00
" Theatre Fires and Panics	12mo,	1 50
Hatfield's American House Carpenter	8vo,	5 00
Holly's Carpenter and Joiner	18mo,	75
Kidder's Architect and Builder's Pocket-book	Morocco flap,	4 00
Merrill's Stones for Building and Decoration	8vo,	5 00
Monckton's Stair Building—Wood, Iron, and Stone	4to,	4 00
Stevens' House Painting	18mo,	75
Wait's Engineering and Architectural Jurisprudence.	*(In the press.)*	
Worcester's Small Hospitals—Establishment and Maintenance, including Atkinson's Suggestions for Hospital Architecture	12mo,	1 25
World's Columbian Exposition of 1893	4to,	2 50

ARMY, NAVY, Etc.

MILITARY ENGINEERING—ORDNANCE—PORT CHARGES, ETC.

Bourne's Screw Propellers	4to,	5 00
Bruff's Ordnance and Gunnery	8vo,	6 00
Bucknill's Submarine Mines and Torpedoes	8vo,	4 00
Chase's Screw Propellers	8vo,	3 00
Cooke's Naval Ordnance	8vo,	12 50
Cronkhite's Gunnery for Non-com. Officers	18mo, morocco,	2 00
De Brack's Cavalry Outpost Duties. (Carr.)	18mo, morocco,	2 00
Dietz's Soldier's First Aid	12mo, morocco,	1 25
* Dredge's Modern French Artillery	4to, half morocco,	20 00
" Record of the Transportation Exhibits Building, World's Columbian Exposition of 1893	4to, half morocco,	15 00
Dyer's Light Artillery	12mo,	3 00
Hoff's Naval Tactics	8vo,	1 50
Hunter's Port Charges	8vo, half morocco,	13 00
Ingalls's Ballistic Tables	8vo,	1 50
" Handbook of Problems in Direct Fire	8vo,	4 00
Mahan's Advanced Guard	18mo,	1 50
" Permanent Fortifications. (Mercur.)	8vo, half morocco,	7 50
Mercur's Attack of Fortified Places	12mo,	2 00

Mercur's Elements of the Art of War	8vo,	$4 00
Metcalfe's Ordnance and Gunnery	12mo, with Atlas,	5 00
Phelps's Practical Marine Surveying	8vo,	2 50
Powell's Army Officer's Examiner	12mo,	4 00
Reed's Signal Service		50
Sharpe's Subsisting Armies	18mo, morocco,	1 50
Strauss and Alger's Naval Ordnance and Gunnery		
Todd and Whall's Practical Seamanship	8vo,	7 50
Very's Navies of the World	8vo, half morocco,	3 50
Wheeler's Siege Operations	8vo,	2 00
Winthrop's Abridgment of Military Law	12mo,	2 50
Woodhull's Notes on Military Hygiene	12mo, morocco,	2 50
Young's Simple Elements of Navigation	12mo, morocco flaps,	2 50

ASSAYING.

Smelting—Ore Dressing—Alloys, Etc.

Fletcher's Quant. Assaying with the Blowpipe	12mo, morocco,	1 50
Furman's Practical Assaying	8vo,	3 00
Kunhardt's Ore Dressing	8vo,	1 50
* Mitchell's Practical Assaying. (Crookes.)	8vo,	10 00
O'Driscoll's Treatment of Gold Ores	8vo,	2 00
Ricketts and Miller's Notes on Assaying	8vo,	3 00
Thurston's Alloys, Brasses, and Bronzes	8vo,	2 50
Wilson's Cyanide Processes	12mo,	1 50
" The Chlorination Process	12mo,	1 50

ASTRONOMY.

Practical, Theoretical, and Descriptive.

Craig's Azimuth	4to,	3 50
Doolittle's Practical Astronomy	8vo,	4 00
Gore's Elements of Geodesy	8vo,	2 50
Michie and Harlow's Practical Astronomy	8vo,	3 00
White's Theoretical and Descriptive Astronomy	12mo,	2 00

BOTANY.

Gardening for Ladies, Etc.

Baldwin's Orchids of New England	8vo,	1 50
Loudon's Gardening for Ladies. (Downing.)	12mo,	1 50
Thomé's Structural Botany	18mo,	2 25
Westermaier's General Botany. (Schneider.)	8vo,	2 00

BRIDGES, ROOFS, Etc.

CANTILEVER—DRAW—HIGHWAY—SUSPENSION.

(*See also* ENGINEERING, p. 6.)

Boller's Highway Bridges	8vo,	$2 00
* " The Thames River Bridge	4to, paper,	5 00
Burr's Stresses in Bridges	8vo,	3 50
Crehore's Mechanics of the Girder	8vo,	5 00
Dredge's Thames Bridges	7 parts,	
Du Bois's Stresses in Framed Structures	4to,	10 00
Foster's Wooden Trestle Bridges	4to,	5 00
Greene's Arches in Wood, etc	8vo,	2 50
" Bridge Trusses	8vo,	2 50
" Roof Trusses	8vo,	1 25
Howe's Treatise on Arches	8vo,	4 00
Johnson's Modern Framed Structures	4to,	10 00
Merriman & Jacoby's Text-book of Roofs and Bridges. Part I., Stresses	8vo,	2 50
Merriman & Jacoby's Text-book of Roofs and Bridges. Part II., Graphic Statics	8vo,	2 50
Merriman & Jacoby's Text-book of Roofs and Bridges. Part III., Bridge Design	8vo,	5 00
Merriman & Jacoby's Text-book of Roofs and Bridges. Part IV., Continuous, Draw, Cantilever, Suspension, and Arched Bridges	(*In preparation*).	
* Morison's The Memphis Bridge	Oblong 4to,	10 00
Waddell's Iron Highway Bridges	8vo,	4 00
Wood's Construction of Bridges and Roofs	8vo,	2 00
Wright's Designing of Draw Spans	8vo,	

CHEMISTRY.

QUALITATIVE—QUANTITATIVE—ORGANIC—INORGANIC, ETC.

Adriance's Laboratory Calculations	12mo,	1 25
Allen's Tables for Iron Analysis	8vo,	3 00
Austen's Notes for Chemical Students	12mo,	1 50
Bolton's Student's Guide in Quantitative Analysis	8vo,	1 50
Classen's Analysis by Electrolysis. (Herrick.)	8vo,	3 00
Crafts's Qualitative Analysis. (Schaeffer.)	12mo,	1 50
Drechsel's Chemical Reactions. (Merrill.)	12mo,	1 25
Fresenius's Quantitative Chemical Analysis. (Allen.)	8vo,	6 00

Fresenius's Qualitative Chemical Analysis. (Johnson.).....8vo,	$4 00
Fuerte's Water and Public Health.......................12mo,	1 50
Gill's Gas and Fuel Analysis...........................12mo,	1 25
Hammarsten's Physiological Chemistry. (Mandel.)........8vo,	4 00
Helm's Principles of Mathematical Chemistry. (Morgan).12mo,	1 50
Kolbe's Inorganic Chemistry...........................12mo,	1 50
Landauer's Spectral Analysis. (Tingle.).........(*In the press*).	
Mandel's Bio-chemical Laboratory......................12mo,	1 50
Mason's Water-supply...................................8vo,	5 00
Miller's Chemical Physics..............................8vo,	2 00
Mixter's Elementary Text-book of Chemistry............12mo,	1 50
Morgan's The Theory of Solutions and its Results.......12mo,	1 00
Nichols's Water Supply (Chemical and Sanitary)..........8vo,	2 50
O'Brine's Laboratory Guide to Chemical Analysis.........8vo,	2 00
Perkins's Qualitative Analysis........................12mo,	1 00
Pinner's Organic Chemistry. (Austen.).................12mo,	1 50
Ricketts and Russell's Notes on Inorganic Chemistry (Non-metallic)........Oblong 8vo, morocco,	75
Schimpf's Volumetric Analysis.........................12mo,	2 50
Spencer's Sugar Manufacturer's Handbook.12mo, morocco flaps,	2 00
" Handbook for Chemists of Beet Sugar House. (*In the press*).	
Stockbridge's Rocks and Soils..........................8vo,	2 50
Troilius's Chemistry of Iron...........................8vo,	2 00
Wiechmann's Chemical Lecture Notes....................12mo,	3 00
" Sugar Analysis.......................................8vo,	2 50
Wulling's Inorganic Phar. and Med. Chemistry..........12mo,	2 00

DRAWING.

Elementary—Geometrical—Topographical.

Hill's Shades and Shadows and Perspective..............8vo,	2 00
MacCord's Descriptive Geometry.........................8vo,	3 00
" Kinematics...8vo,	5 00
" Mechanical Drawing...................................8vo,	4 00
Mahan's Industrial Drawing. (Thompson.)........2 vols., 8vo,	3 50
Reed's Topographical Drawing. (H. A.)..................4to,	5 00
Smith's Topographical Drawing. (Macmillan.)............8vo,	2 50
Warren's Descriptive Geometry..................2 vols., 8vo,	3 50
" Drafting Instruments................................12mo,	1 25

Warren's Free-hand Drawing	12mo,	$1 00
" Higher Linear Perspective	8vo,	3 50
" Linear Perspective	12mo,	1 00
" Machine Construction	2 vols., 8vo,	7 50
" Plane Problems	12mo,	1 25
" Primary Geometry	12mo,	75
" Problems and Theorems	8vo,	2 50
" Projection Drawing	12mo,	1 50
" Shades and Shadows	8vo,	3 00
" Stereotomy—Stone Cutting	8vo,	2 50
Whelpley's Letter Engraving	12mo,	2 00

ELECTRICITY AND MAGNETISM.

ILLUMINATION—BATTERIES—PHYSICS.

Anthony and Brackett's Text-book of Physics (Magie).	8vo,	4 00
Barker's Deep-sea Soundings	8vo,	2 00
Benjamin's Voltaic Cell	8vo,	3 00
Cosmic Law of Thermal Repulsion	18mo,	75
Crehore and Squier's Experiments with a New Polarizing Photo-Chronograph	8vo,	3 00
* Dredge's Electric Illuminations	2 vols., 4to, half morocco,	25 00
" " " Vol. II	4to,	7 50
Gilbert's De magnete. (Mottelay.)	8vo,	2 50
Holman's Precision of Measurements	8vo,	2 00
Michie's Wave Motion Relating to Sound and Light,	8vo,	4 00
Morgan's, The Theory of Solutions and its Results	12mo,	1 00
Niaudet's Electric Batteries. (Fishback.)	12mo,	2 50
Reagan's Steam and Electrical Locomotives	12mo	2 00
Thurston's Stationary Steam Engines for Electric Lighting Purposes	12mo,	1 50
Tillman's Heat	8vo,	1 50

ENGINEERING.

CIVIL—MECHANICAL—SANITARY, ETC.

(*See also* BRIDGES, p. 4; HYDRAULICS, p. 8; MATERIALS OF ENGINEERING, p. 9; MECHANICS AND MACHINERY, p. 11; STEAM ENGINES AND BOILERS, p. 14.)

Baker's Masonry Construction	8vo,	5 00
Baker's Surveying Instruments	12mo,	3 00
Black's U. S. Public Works	4to,	5 00

Butts's Engineer's Field-book................12mo, morocco,		$2 50
Byrne's Highway Construction..........................8vo,		7 50
Carpenter's Experimental Engineering 8vo,		6 00
Church's Mechanics of Engineering—Solids and Fluids....8vo,		6 00
" Notes and Examples in Mechanics...............8vo,		2 00
Crandall's Earthwork Tables 8vo,		1 50
Crandall's The Transition Curve...............12mo, morocco,		1 50
* Dredge's Penn. Railroad Construction, etc... Folio, half mor.,		20 00
* Drinker's Tunnelling...................4to, half morocco,		25 00
Eissler's Explosives—Nitroglycerine and Dynamite........8vo,		4 00
Gerhard's Sanitary House Inspection....................16mo,		1 00
Godwin's Railroad Engineer's Field-book.12mo, pocket-bk. form,		2 50
Gore's Elements of Goodesy......................8vo,		2 50
Howard's Transition Curve Field-book.....12mo, morocco flap,		1 50
Howe's Retaining Walls (New Edition.)..................12mo,		1 25
Hudson's Excavation Tables. Vol. II....................8vo,		1 00
Hutton's Mechanical Engineering of Power Plants........8vo,		5 00
Johnson's Materials of Construction......................8vo,		6 00
Johnson's Stadia Reduction Diagram..Sheet, $22\frac{1}{2} \times 28\frac{1}{2}$ inches,		50
" Theory and Practice of Surveying...............8vo,		4 00
Kent's Mechanical Engineer's Pocket-book.....12mo, morocco,		5 00
Kiersted's Sewage Disposal..........................12mo,		1 25
Kirkwood's Lead Pipe for Service Pipe..................8vo,		1 50
Mahan's Civil Engineering. (Wood.)....................8vo,		5 00
Merriman and Brook's Handbook for Surveyors....12mo, mor.,		2 00
Merriman's Geodetic Surveying..........................8vo,		2 00
" Retaining Walls and Masonry Dams..........8vo,		2 00
Mosely's Mechanical Engineering. (Mahan.)............8vo,		5 00
Nagle's Manual for Railroad Engineers........12mo, morocco,		
Patton's Civil Engineering.............................8vo,		7 50
" Foundations....................................8vo,		5 00
Rockwell's Roads and Pavements in France............12mo,		1 25
Ruffner's Non-tidal Rivers 8vo,		1 25
Searles's Field Engineering...............12mo, morocco flaps,		3 00
Searles's Railroad Spiral ....12mo, morocco flaps,		1 50
Siebert and Biggin's Modern Stone Cutting and Masonry...8vo,		1 50
Smith's Cable Tramways.................................4to,		2 50
" Wire Manufacture and Uses........................4to,		3 00

Spalding's Roads and Pavements.......................12mo,		$2 00
" Hydraulic Cement...........................12mo,		2 00
Thurston's Materials of Construction 8vo,		5 00
* Trautwine's Civil Engineer's Pocket-book...12mo, mor. flaps,		5 00
* " Cross-section................................Sheet,		25
* " Excavations and Embankments.............8vo,		2 00
* " Laying Out Curves.............12mo, morocco,		2 50
Wait's Engineering and Architectural Jurisprudence.		
(*In the press.*)		
Warren's Stereotomy—Stone Cutting.........................8vo,		2 50
Webb's Engineering Instruments.............12mo, morocco,		1 00
Wegmann's Construction of Masonry Dams...............4to,		5 00
Wellington's Location of Railways.............................8vo,		5 00
Wheeler's Civil Engineering......................................8vo,		4 00
Wolff's Windmill as a Prime Mover.........................8vo,		3 00

HYDRAULICS.

WATER-WHEELS—WINDMILLS—SERVICE PIPE—DRAINAGE, ETC.

(*See also* ENGINEERING, p. 6.)

Bazin's Experiments upon the Contraction of the Liquid Vein (Trautwine)..8vo,	2 00
Bovey's Treatise on Hydraulics..................................8vo,	4 00
Coffin's Graphical Solution of Hydraulic Problems.........12mo,	2 50
Ferrel's Treatise on the Winds, Cyclones, and Tornadoes...8vo,	4 00
Fuerte's Water and Public Health.........................12mo,	1 50
Ganguillet & Kutter's Flow of Water. (Hering & Trautwine.) 8vo,	4 00
Hazen's Filtration of Public Water Supply.................8vo,	2 00
Herschel's 115 Experiments..................................8vo,	2 00
Kiersted's Sewage Disposal...................................12mo,	1 25
Kirkwood's Lead Pipe for Service Pipe.....................8vo,	1 50
Mason's Water Supply..8vo,	5 00
Merriman's Treatise on Hydraulics...........................8vo,	4 00
Nichols's Water Supply (Chemical and Sanitary)...........8vo,	2 50
Ruffner's Improvement for Non-tidal Rivers................8vo,	1 25
Wegmann's Water Supply of the City of New York 4to,	10 00
Weisbach's Hydraulics. (Du Bois.)..........................8vo,	5 00
Wilson's Irrigation Engineering 8vo,	4 00
Wolff's Windmill as a Prime Mover.........................8vo,	3 00
Wood's Theory of Turbines....................................8vo,	2 50

MANUFACTURES.

ANILINE—BOILERS—EXPLOSIVES—IRON—SUGAR—WATCHES—WOOLLENS, ETC.

Allen's Tables for Iron Analysis	8vo,	$3 00
Beaumont's Woollen and Worsted Manufacture	12mo,	1 50
Bolland's Encyclopædia of Founding Terms	12mo,	3 00
" The Iron Founder	12mo,	2 50
" " " " Supplement	12mo,	2 50
Booth's Clock and Watch Maker's Manual	12mo,	2 00
Bouvier's Handbook on Oil Painting	12mo,	2 00
Eissler's Explosives, Nitroglycerine and Dynamite	8vo,	4 00
Ford's Boiler Making for Boiler Makers	18mo,	1 00
Metcalfe's Cost of Manufactures	8vo,	5 00
Metcalf's Steel—A Manual for Steel Users	12mo,	2 00
Reimann's Aniline Colors. (Crookes.)	8vo,	2 50
* Reisig's Guide to Piece Dyeing	8vo,	25 00
Spencer's Sugar Manufacturer's Handbook	12mo, mor. flap,	2 00
" Handbook for Chemists of Beet Houses. (*In the press.*)		
Svedelius's Handbook for Charcoal Burners	12mo,	1 50
The Lathe and Its Uses	8vo,	6 00
Thurston's Manual of Steam Boilers	8vo,	5 00
Walke's Lectures on Explosives	8vo,	4 00
West's American Foundry Practice	12mo,	2 50
" Moulder's Text-book	12mo,	2 50
Wiechmann's Sugar Analysis	8vo,	2 50
Woodbury's Fire Protection of Mills	8vo,	2 50

MATERIALS OF ENGINEERING.

STRENGTH—ELASTICITY—RESISTANCE, ETC.

(*See also* ENGINEERING, p. 6.)

Baker's Masonry Construction	8vo,	5 00
Beardslee and Kent's Strength of Wrought Iron	8vo,	1 50
Bovey's Strength of Materials	8vo,	7 50
Burr's Elasticity and Resistance of Materials	8vo,	5 00
Byrne's Highway Construction	8vo,	5 00
Carpenter's Testing Machines and Methods of Testing Materials		
Church's Mechanic's of Engineering—Solids and Fluids	8vo,	6 00
Du Bois's Stresses in Framed Structures	4to,	10 00

Hatfield's Transverse Strains........................ 8vo, $5 00
Johnson's Materials of Construction......................8vo, 6 00
Lanza's Applied Mechanics.............................8vo, 7 50
" Strength of Wooden Columns.............8vo, paper, 50
Merrill's Stones for Building and Decoration..............8vo, 5 00
Merriman's Mechanics of Materials.......................8vo, 4 00
Patton's Treatise on Foundations........................8vo, 5 00
Rockwell's Roads and Pavements in France.............12mo, 1 25
Spalding's Roads and Pavements......................12mo, 2 00
Thurston's Materials of Construction...................8vo, 5 00
Thurston's Materials of Engineering...............3 vols., 8vo, 8 00
Vol. I., Non-metallic...............................8vo, 2 00
Vol. II., Iron and Steel...........................8vo, 3 50
Vol. III., Alloys, Brasses, and Bronzes...............8vo, 2 50
Weyrauch's Strength of Iron and Steel. (Du Bois.)........8vo, 1 50
Wood's Resistance of Materials..........................8vo, 2 00

MATHEMATICS.

Calculus—Geometry—Trigonometry, Etc.

Baker's Elliptic Functions...............................8vo, 1 50
Ballard's Pyramid Problem..............................8vo, 1 50
Barnard's Pyramid Problem..............................8vo, 1 50
Bass's Differential Calculus...........................12mo, 4 00
Brigg's Plane Analytical Geometry......................12mo, 1 00
Chapman's Theory of Equations.........................12mo, 1 50
Chessin's Elements of the Theory of Functions...............
Compton's Logarithmic Computations...................12mo, 1 50
Craig's Linear Differential Equations....................8vo, 5 00
Davis's Introduction to the Logic of Algebra..............8vo, 1 50
Halsted's Elements of Geometry.........................8vo, 1 75
" Synthetic Geometry...........................8vo, 1 50
Johnson's Curve Tracing..............................12mo, 1 00
" Differential Equations—Ordinary and Partial.....8vo, 3 50
" Integral Calculus..........................12mo, 1 50
" Least Squares.............................12mo, 1 50
Ludlow's Logarithmic and Other Tables. (Bass.).........8vo, 2 00
" Trigonometry with Tables. (Bass.).............8vo, 3 00
Mahan's Descriptive Geometry (Stone Cutting)............8vo, 1 50

Merriman and Woodward's Higher Mathematics	8vo,	$5 00
Merriman's Method of Least Squares	8vo,	2 00
Parker's Quadrature of the Circle	8vo,	2 50
Rice and Johnson's Differential and Integral Calculus,	2 vols. in 1, 12mo,	2 50
" Differential Calculus	8vo,	3 50
" Abridgment of Differential Calculus	8vo,	1 50
Searles's Elements of Geometry	8vo,	1 50
Totten's Metrology	8vo,	2 50
Warren's Descriptive Geometry	2 vols., 8vo,	3 50
" Drafting Instruments	12mo,	1 25
" Free-hand Drawing	12mo,	1 00
" Higher Linear Perspective	8vo,	3 50
" Linear Perspective	12mo,	1 00
" Primary Geometry	12mo,	75
" Plane Problems	12mo,	1 25
" Plane Problems	12mo,	1 25
" Problems and Theorems	8vo,	2 50
" Projection Drawing	12mo,	1 50
Wood's Co-ordinate Geometry	8vo,	2 00
" Trigonometry	12mo,	1 00
Woolf's Descriptive Geometry	Royal 8vo,	3 00

MECHANICS—MACHINERY.

TEXT-BOOKS AND PRACTICAL WORKS.

(*See also* ENGINEERING, p. 6.)

Baldwin's Steam Heating for Buildings	12mo,	2 50
Benjamin's Wrinkles and Recipes	12mo,	2 00
Carpenter's Testing Machines and Methods of Testing Materials	8vo,	
Chordal's Letters to Mechanics	12mo,	2 00
Church's Mechanics of Engineering	8vo,	6 00
" Notes and Examples in Mechanics	8vo,	2 00
Crehore's Mechanics of the Girder	8vo,	5 00
Cromwell's Belts and Pulleys	12mo,	1 50
" Toothed Gearing	12mo,	1 50
Compton's First Lessons in Metal Working	12mo,	1 50
Dana's Elementary Mechanics	12mo,	1 50

Dingey's Machinery Pattern Making	12mo,	$2 00
Dredge's Trans. Exhibits Building, World Exposition,	4to, half morocco,	15 00
Du Bois's Mechanics. Vol. I., Kinematics	8vo,	3 50
" " Vol. II., Statics	8vo,	4 00
" " Vol. III., Kinetics	8vo,	3 50
Fitzgerald's Boston Machinist	18mo,	1 00
Flather's Dynamometers	12mo,	2 00
" Rope Driving	12mo,	2 00
Hall's Car Lubrication	12mo,	1 00
Holly's Saw Filing	18mo,	75
Lanza's Applied Mechanics	8vo,	7 50
MacCord's Kinematics	8vo,	5 00
Merriman's Mechanics of Materials	8vo,	4 00
Metcalfe's Cost of Manufactures	8vo,	5 00
Michie's Analytical Mechanics	8vo,	4 00
Mosely's Mechanical Engineering. (Mahan.)	8vo,	5 00
Richards's Compressed Air	12mo,	1 50
Robinson's Principles of Mechanism	8vo,	3 00
Smith's Press-working of Metals	8vo,	3 00
The Lathe and Its Uses	8vo,	6 00
Thurston's Friction and Lost Work	8vo,	3 00
" The Animal as a Machine	12mo,	1 00
Warren's Machine Construction	2 vols., 8vo,	7 50
Weisbach's Hydraulics and Hydraulic Motors. (Du Bois.)	8vo,	5 00
" Mechanics of Engineering. Vol. III., Part I., Sec. I. (Klein.)	8vo,	5 00
Weisbach's Mechanics of Engineering Vol. III., Part I., Sec. II. (Klein.)	8vo,	5 00
Weisbach's Steam Engines. (Du Bois.)	8vo,	5 00
Wood's Analytical Mechanics	8vo,	3 00
" Elementary Mechanics	12mo,	1 25
" " " Supplement and Key		1 25

METALLURGY.

IRON—GOLD—SILVER—ALLOYS, ETC.

Allen's Tables for Iron Analysis	8vo,	3 00
Egleston's Gold and Mercury	8vo,	7 50

Egleston's Metallurgy of Silver 8vo, $7 50
* Kerl's Metallurgy—Copper and Iron.....................8vo, 15 00
* " " Steel, Fuel, etc......................8vo, 15 00
Kunhardt's Ore Dressing in Europe.......................8vo, 1 50
Metcalf Steel—A Manual for Steel Users... ....12mo, 2 00
O'Driscoll's Treatment of Gold Ores.....................8vo, 2 00
Thurston's Iron and Steel...............................8vo, 3 50
" Alloys...8vo, 2 50
Wilson's Cyanide Processes....... ....12mo, 1 50

MINERALOGY AND MINING.

Mine Accidents—Ventilation—Ore Dressing, Etc.

Barringer's Minerals of Commercial Value........(*In the pre s.*)
Beard's Ventilation of Mines............... ....12mo, 2 50
Boyd's Resources of South Western Virginia..............8vo, 3 00
" Map of South Western Virginia.....Pocket-book form, 2 00
Brush and Penfield's Determinative Mineralogy.. ....8vo, 3 50
Chester's Catalogue of Minerals.........................8vo, 1 25
" Dictionary of the Names of Minerals...................8vo, 3 00
Dana's American Localities of Minerals..................8vo, 1 00
" Descriptive Mineralogy. (E. S.).....8vo, half morocco, 12 50
" Mineralogy and Petrography. (J. D.)...........12mo, 2 00
" Minerals and How to Study Them. (E. S.)......12mo, 1 50
" Text-book of Mineralogy. (E. S.)................8vo, 3 50
*Drinker's Tunnelling, Explosives, Compounds, and Rock Drills.
4to, half morocco, 25 00
Egleston's Catalogue of Minerals and Synonyms...........8vo, 2 50
Eissler's Explosives—Nitroglycerine and Dynamite........8vo, 4 00
Goodyear's Coal Mines of the Western Coast............12mo, 2 50
Hussak's Rock-forming Minerals. (Smith.)...............8vo, 2 00
Ihlseng's Manual of Mining.. ....8vo, 4 00
Kunhardt's Ore Dressing in Europe.......................8vo, 1 50
O'Driscoll's Treatment of Gold Ores.....................8vo, 2 00
Rosenbusch's Microscopical Physiography of Minerals and
Rocks. (Iddings.).......................................8vo, 5 00
Sawyer's Accidents in Mines............. ....8vo, 7 00
Stockbridge's Rocks and Soils...........................8vo, 2 50

Williams's Lithology................................8vo, $3 00
Wilson's Mine Ventilation............................16mo, 1 25

STEAM AND ELECTRICAL ENGINES, BOILERS, Etc.

STATIONARY—MARINE—LOCOMOTIVE—GAS ENGINES, ETC.

(*See also* ENGINEERING, p. 6.)

Baldwin's Steam Heating for Buildings.................12mo, 2 50
Clerk's Gas Engine 12mo, 4 00
Ford's Boiler Making for Boiler Makers................18mo, 1 00
Hemenway's Indicator Practice.........................12mo, 2 00
Hoadley's Warm-blast Furnace...........................8vo, 1 50
Kneass's Practice and Theory of the Injector 8vo, 1 50
MacCord's Slide Valve..................................8vo,
* Maw's Marine Engines.................Folio, half morocco, 18 00
Meyer's Modern Locomotive Construction.................4to, 10 00
Peabody and Miller's Steam Boilers.....................8vo, 4 00
Peabody's Tables of Saturated Steam....................8vo, 1 00
" Thermodynamics of the Steam Engine......... 8vo, 5 00
" Valve Gears for the Steam-Engine.............8vo, 2 50
Pray's Twenty Years with the Indicator............Royal 8vo, 2 50
Pupin and Osterberg's Thermodynamics..................12mo, 1 25
Reagan's Steam and Electrical Locomotives.............12mo, 2 00
Röntgen's Thermodynamics. (Du Bois.)...................8vo, 5 00
Sinclair's Locomotive Running.........................12mo, 2 00
Thurston's Boiler Explosion...........................12mo, 1 50
" Engine and Boiler Trials.........................8vo, 5 00
" Manual of the Steam Engine. Part I., Structure and Theory 8vo, 7 50
" Manual of the Steam Engine. Part II., Design, Construction, and Operation...............8vo, 7 50
2 parts, 12 00
" Philosophy of the Steam Engine.............12mo, 75
" Reflection on the Motive Power of Heat. (Carnot.) 12mo, 2 00
" Stationary Steam Engines..................12mo, 1 50
" Steam-boiler Construction and Operation.......8vo, 5 00
Spangler's Valve Gears.................................8vo, 2 50

Trowbridge's Stationary Steam Engines	4to, boards,	$2 50
Weisbach's Steam Engine. (Du Bois.)	8vo,	5 00
Whitham's Constructive Steam Engineering	8vo,	10 00
" Steam-engine Design	8vo,	6 00
Wilson's Steam Boilers. (Flather.)	12mo,	2 50
Wood's Thermodynamics, Heat Motors, etc.	8vo,	4 00

TABLES, WEIGHTS, AND MEASURES.

For Actuaries, Chemists, Engineers, Mechanics—Metric Tables, Etc.

Adriance's Laboratory Calculations	12mo,	1 25
Allen's Tables for Iron Analysis	8vo,	3 00
Bixby's Graphical Computing Tables	Sheet,	25
Compton's Logarithms	12mo,	1 50
Crandall's Railway and Earthwork Tables	8vo,	1 50
Egleston's Weights and Measures	18mo,	75
Fisher's Table of Cubic Yards	Cardboard,	25
Hudson's Excavation Tables. Vol. II	8vo,	1 00
Johnson's Stadia and Earthwork Tables	8vo,	1 25
Ludlow's Logarithmic and Other Tables. (Bass.)	12mo,	2 00
Thurston's Conversion Tables	8vo,	1 00
Totten's Metrology	8vo,	2 50

VENTILATION.

Steam Heating—House Inspection—Mine Ventilation.

Baldwin's Steam Heating	12mo,	2 50
Beard's Ventilation of Mines	12mo,	2 50
Carpenter's Heating and Ventilating of Buildings	8vo,	3 00
Gerhard's Sanitary House Inspection	Square 16mo,	1 00
Mott's The Air We Breathe, and Ventilation	16mo,	1 00
Reid's Ventilation of American Dwellings	12mo,	1 50
Wilson's Mine Ventilation	16mo,	1 25

MISCELLANEOUS PUBLICATIONS.

Alcott's Gems, Sentiment, Language	Gilt edges,	5 00
Bailey's The New Tale of a Tub	8vo,	75
Ballard's Solution of the Pyramid Problem	8vo,	1 50
Barnard's The Metrological System of the Great Pyramid	8vo,	1 50

Davis' Elements of Law	8vo,	$2 00
Emmon's Geological Guide-book of the Rocky Mountains	8vo,	1 50
Ferrel's Treatise on the Winds	8vo,	4 00
Haines' Addresses Delivered before the Am. Ry. Assn.	12mo. (*In the press.*)	
Mott's The Fallacy of the Present Theory of Sound	Sq. 16mo,	1 00
Perkins's Cornell University	Oblong 4to,	1 50
Ricketts's History of Rensselaer Polytechnic Institute	8vo,	3 00
Rotherham's The New Testament Critically Emphasized.	12mo,	1 50
Totten's An Important Question in Metrology	8vo,	2 50
Whitehouse's Lake Mœris	Paper,	25
* Wiley's Yosemite, Alaska, and Yellowstone	4to,	3 00

HEBREW AND CHALDEE TEXT-BOOKS.

For Schools and Theological Seminaries.

Gesenius's Hebrew and Chaldee Lexicon to Old Testament. (Tregelles.)	Small 4to, half morocco,	5 00
Green's Elementary Hebrew Grammar	12mo,	1 25
" Grammar of the Hebrew Language (New Edition)	8vo,	3 00
" Hebrew Chrestomathy	8vo,	2 00
Letteris's Hebrew Bible (Massoretic Notes in English).	8vo, arabesque,	2 25
Luzzato's Grammar of the Biblical Chaldaic Language and the Talmud Babli Idioms	12mo,	1 50

MEDICAL.

Bull's Maternal Management in Health and Disease	12mo,	1 00
Hammarsten's Physiological Chemistry. (Mandel.)	8vo,	4 00
Mott's Composition, Digestibility, and Nutritive Value of Food.	Large mounted chart,	1 25
Ruddiman's Incompatibilities in Prescriptions	(*In the press.*)	
Steel's Treatise on the Diseases of the Ox	8vo,	3 00
" Treatise on the Diseases of the Dog	8vo,	3 50
Worcester's Small Hospitals—Establishment and Maintenance, including Atkinson's Suggestions for Hospital Architecture	12mo,	1 25